50 Years of Language Experiments with Great Apes

Igor Hanzel

50 Years of Language Experiments with Great Apes

PETER LANG
EDITION

Bibliographic Information published by the Deutsche Nationalbibliothek
The Deutsche Nationalbibliothek lists this publication in the Deutsche Nationalbibliografie; detailed bibliographic data is available in the internet at http://dnb.d-nb.de.

Library of Congress Cataloging-in-Publication Data:
Names: Hanzel, Igor.
Title: 50 years of language experiments with great apes / Igor Hanzel.
Other titles: Fifty years of language experiments with great apes
Description: Frankfurt am Main : Peter Lang GmbH, 2017. | Includes bibliographical references.
Identifiers: LCCN 2017006397 | ISBN 9783631720936
Subjects: LCSH: Hominids—Behavior. | Apes—Behavior. | Animal communication. | Human-animal communication.
Classification: LCC QL776 .H35 2017 | DDC 591.59/4—dc23 LC record available at https://lccn.loc.gov/2017006397

ISBN 978-3-631-72093-6 (Print)
E-ISBN 978-3-631-72090-5 (E-PDF)
E-ISBN 978-3-631-72091-2 (EPUB)
E-ISBN 978-3-631-72092-9 (MOBI)
DOI 10.3726/b11008

© Peter Lang GmbH
Internationaler Verlag der Wissenschaften
Frankfurt am Main 2017
All rights reserved.
Peter Lang Edition is an Imprint of Peter Lang GmbH.

Peter Lang – Frankfurt am Main · Bern · Bruxelles · New York · Oxford · Warszawa · Wien

This publication has been peer reviewed.

www.peterlang.com

Abstract

The book approaches the language experiments with great apes performed in the last 50 years from the point of view of logical semantics, speech-act theory, and philosophy of the social sciences based on the linguistic turn in philosophy. Subjected to analysis are the experiments organized by D. Premack, D. M. Rumbaugh, E. S. Savage-Rumbaugh, R. A. Gardner and B. T. Gardner, L. W. Miles and S. Boysen. None of these experiments involved a thoroughly developed understanding of what language stands for, instead, the scientists participating in those experiments tried to bypass the delineation of the notion of language by introducing some surrogates for it, taken from the behavioristic tradition or from some form of tentatively delineated cognitivism. These attempts are symptomatic of an unfinished turn to language in language experiments with great apes. The book shows that once this turn is accomplished by employing the understanding of the notion of language developed in modern logical semantics and speech-act theory, then past language experiments with great apes can be reevaluated and new language experiments can be proposed.

Abstract

Contents

Acknowledgments

This work owes very much to the discussions I had with Professor H.-P. Krüger at the Institute of Philosophy at the University of Potsdam, Potsdam, Germany in summer 2015. Special thanks go to D. Glavaničová, E. S. Savage-Rumbaugh, R. A. Gardner and S. T. Boysen for their helpful insights and patient advices.

I am grateful to the Springer publisher for the permission to reproduce a table from D. M. Rumbaugh's article "A computer-controlled language training system for investigating the language skills of young apes" page 386, published in the journal *Behavior Research Methods and Instrumentation*, 1973, Volume 5, Issue 5.

John Wiley and Sons gave the permission to reproduce a figure from E. S. Savage-Rumbaugh's article: "Can apes use symbols to represent their world?" pages 46–47, published in the *Annals of the New York Academy of Sciences*, 1981, Volume 364.

The journal *Science* gave the permission to reproduce a figure from D. Premack's article "Language in chimpanzee?" page 809, published in this journal, Volume 172, Issue 3985.

Cambridge University Press gave the permission to reproduce three figures from J. Dore's article "Holophrases, speech acts and language universals," pages 34–35, published in *Journal of Child Language*, 1975, Volume 2, Issue 1.

1. Introduction

The aim of this study is twofold. First, I shall give an overview of the research into the language capabilities of great apes conducted in the last 50 years. This research was initiated in June 1966, when R. A. Gardner and B. T. Gardner started to teach American Sign Language (ASL) to Washoe, a female chimpanzee. I will scrutinize the language experiments based on ASL, the so-called Yerkish Language (henceforth YL), the experiments employing symbols embodied in plastic tokens (henceforth PT), and finally, the experiments targeting the great apes' skill to master the language of mathematics. Second, I shall interpret both the nature of the languages employed in the experiments and the results of the experiments from the point of view of modern logical semantics and elements of speech-act theory.

There are two reasons for choosing this point of view, a more thorough justification of which will be given in Part 7. First, while the experiments with great apes I have mentioned involved three types of languages, none of these experiments was based on a thoroughly developed understanding of *what language stands for*. Instead, the scientists participating in those experiments very often tried to bypass or replace the notion of language by introducing some surrogates for it, taken, for example, from the behavioristic (Skinnerian) tradition or from some form of tentatively delineated cognitivism.

These attempts are symptom of a paradox – namely, of an *unfinished turn to language* in *language* experiments with great apes. As I will try to show, once this turn is accomplished by employing the understanding of the notion of language developed in modern logical semantics and speech-act theory, then not only language experiments with great apes in the past can be evaluated anew, but also new language experiments can be proposed.

Second, my orientation to modern logical semantics and speech-act theory was also prompted by the fact that experimenters involved in language experiments with great apes were well aware that these experiments faced the issue of defining the notion of language. The following quotes from their works bear witness to this:

> The program we have described avoids the question of whether an animal other than man can acquire language. As comparative psychologists we must reject this question ... It depends on the definition of language (Gardner and Gardner 1971, 181).

> Given the chimpanzees' apparent problems with vocal production, the next logical step is to see if they can produce language in another modality. The Gardners and others have attacked this problem. But once this is done, we immediately run into the problem of not knowing what language is in definitional terms ... there is no generally accepted definition of language (Fouts 1978, 130).

> The answer to the question "Is the ape capable of human-type language skills?" clearly pivots on the definition of language (Rumbaugh and Gill 1976, 90).

I start with an overview of three language-in-ape projects – the ASL project, the YL project, the PT project – and with an overview of mathematical experiments with great apes. Then I propose a view of language that draws on the modern logical semantics and the typology of illocutionary acts provided by the theory of speech acts. Based on that view and that typology, I then reevaluate those three projects and the mathematical addition experiment. Next, I delineate the methods employed in those three language-in-ape projects. By way of conclusion and as a result of, I hope, a thoroughly accomplished turn to language, I propose a set of new experiments with great apes.

Before explicating the meaning of the term "language", it is necessary to make the following terminological clarification. Even if in English the term "sign languages" is standardly understood to refer to languages used by the deaf, such an understanding is in fact misleading because languages produced in other modalities – for example vocal modality – also use signs. It is also worth noting that in some languages other than English, languages used by the deaf are labeled by a term referring not to signs but to gestures.[1] For this reason I will use the terms "gesture-signs", "Yerkish signs" and "plastic token signs", respectively, with the corresponding projects; and instead of the standardly used abbreviation ASL, I will use AGSL as an abbreviation for "American Gesture Sign Language".

---

1   For example, in German they are labeled as "Gebärdensprachen" – that is, as "gesture languages".

2. The Gesture-Signs: Washoe and Chantek

2.1 Washoe

The project to teach a gesture-sign language was delineated by the Gardners as follows (1974, 3):

> Project Washoe ... is best understood as a pilot study in a program aimed at establishing a truly comparative psychology of two way communication ... we set out to demonstrate that a chimpanzee could achieve a significant degree of two-way communication by using a genuine form of human language.

The realization of this project was characterized from its beginning by the attempt to bypass any characterization of human language and/or language in general. They gave two reasons for this. First, linguistics and psycholinguistics did not succeed in devising a behavioral definition of language (Gardner and Gardner 1975, 244), and, second, the more technically oriented linguists seemed "to be concerned with ideal performance of theoretical human beings, with little or no acknowledgment of the difference in competence between toddlers and college professors" (Gardner and Gardner 1974, 3). As a consequence, the Gardners tried to avoid any a priori definition of language.

As is evident from the following quotation, it is worth mentioning that the Gardners understood the term "behavioral" in a strongly naturalist (natural science) tradition: "Our research on teaching sign language to chimpanzees proceeds from the assumption that any form of behavior, human, or animal, if it exists at all, exists as a natural, biological phenomenon" (Gardner and Gardner 1978, 37). This, in turn, led them to state their allegiance to the behavioristic stimulus-response (S-R) paradigm (Gardner and Gardner 1971, 129):

15

The acquisition of individual signs is the aspect of this project that is most related to the paradigm of S-R reinforcement theory. This paradigm which had a strong influence on the tactics that we used for teaching individual signs, serves as a convenient point of departure for the description of specific teaching methods. According to this theory, particular responses are made in particular stimulus situations. When a response is followed promptly by a reward, there is an increment in the probability that that particular response will be repeated by the subject when that particular stimulus situation is repeated.

They viewed the framework for teaching Washoe elements of AGSL as realized in the method of shaping. Thus, by means of this method they "could select, from among the responses that Washoe tended to make in a given stimulus condition, a response that resembled a sign in ASL that was appropriate to the situation. By administering suitable rewards we could then shape increasingly closer approximations of the to-be-learned sign" (1971, 132).

In addition to the method of shaping they also employed the methods of molding (guidance), imitation, and free style. The first stood for the guiding of Washoe's hand so that it acquired the shape, position, and movement corresponding to a word in AGSL, the second stood for Washoe's reproduction of a word presented to her by the experimenter, and the third for a blend of the previous two methods (Fouts 1972).[2]

Once these methods are applied, according to the Gardners (1974, 3), the actual linguistic performance of chimpanzees participating in the experiments has to be compared with the performance of human children, and once the former match the latter, then one could declare that great apes have

---

2   For details of these methods see (Fouts 1975). On the method of molding applied to other AGSL-taught chimpanzees, see (Fouts 1973).

acquired a language used by humans. Let me now turn to the results achieved by experiments with chimpanzees performed by the Gardners, the Fouts, and their collaborators.

The gesture-sign vocabulary of the chimpanzees, in the course of being taught AGSL, underwent a rapid expansion. So, for example, Washoe in 51 months mastered a vocabulary of 132 gesture-signs (Gardner and Van Cantfort and Gardner 1992) and after 10 years possessed a vocabulary of 200 gesture-signs (Fouts and Shapiro and O'Niel 1978); she was able to apply these gesture-signs to new exemplars of their respective referents (Gardner and Gardner 1969).[3]

In addition to the ability to apply a sign to new exemplars of its referent, Washoe was also able to, after 4 years of training, combine in a creative – that is, not previously taught – manner, two, three, or four of them into sentences; Table 1 lists the latter (Brown 1980, 86)[4]:

Table 1 Examples of Washoe's sequence of gesture-signs

> A. Two Signs
> > 1. Using "emphasizers" *(please, come-gimme, hurry, more)*
> > "HURRY OPEN"
> > "MORE SWEET"
> > "MORE TICKLE"
> > "COME-GIMME DRINK"
> > 2. Using "specifiers"
> > "GO SWEET" (to be carried to fruitbushes).
> > "LISTEN EAT" (at sound of supper bell).
> > "LISTEN DOG" (at sound of barking).

---

3 For the dynamics of vocabulary of other chimpanzees participating in the AGSL project, see (B. Gardner and Gardner 1998).

4 I reproduce all expressions produced in AGSL, YL, and PT projects in capital letters placed in quotation marks.

3. Using names or pronouns
   "YOU DRINK"
   "YOU EAT"
   "ROGER COME"

B. Three or More Signs
   1. Using "emphasizers"
      "GIMME PLEASE FOOD"
      "PLEASE TICKLE MORE"
      "HURRY GIMME TOOTHBRUSH"
   2. Using "specifiers"
      "KEY OPEN FOOD"
      "OPEN KEY CLEAN"
      "KEY OPEN PLEASE BLANKET"
   3. Using names or pronouns
      "YOU ME GO-THERE IN"
      "YOU OUT GO"
      "ROGER WASHOE TICKLE"

Two episodes of Washoe's creativity pertain also to her ability to innovatively combine gesture-signs into a new name of a referent. After being taught the gesture-signs "WATER" and "BIRD", but not the gesture-sign "DUCK", she was still able to label a duck she saw swimming in water by spontaneously combining these two already taught signs into "WATER BIRD". In a similar manner she was able to unify the signs "ROCK" and "BERRY" to create "ROCK BERRY" to refer to a brazil nut (Fouts 1974a).

While the gesture-signs were initially acquired with respect to their referents, which were visible to the chimpanzees, in another experiment (Fouts and Chown and Gooding 1976), a male chimpanzee, already in command of 78 gesture-signs, was able to acquire additional 10 of them without their respective referents being visible to him. He was taught, first, to identify referents of 10 spoken words and then, the 10 gesture-signs corresponding to those spoken words, while the respective referents were not present to the chimpanzee. The test of his mastery of these 10 gesture-signs consisted in confronting him with the referents of these 10 gesture-signs while being asked in AGSL: "WHAT IS THAT?". His success rate was 100 %.

While all the cases of Washoe's mastery of gesture-signs mentioned earlier were somehow related to interaction with human experimenters, it is worth noting that the AGSL project aimed quite early at a feature of language interaction among humans – namely, in the course of interaction in the medium of language, language is reproduced; that is, via use, language becomes a self-sustaining entity (Fouts 1974b; Fouts and Shapiro and O'Niel 1978; Fouts and Hirsch and Fouts 1982). The existence of this self-sustainability of AGSL among chimpanzees was discovered when Washoe started spontaneously to teach gesture-signs to his adopted infant Loulis by directly presenting to him the pair <gesture-sign, referent> and applying often the method of molding the infant's hands into the right form. The result of these procedures was that Loulis, by the age of 29 months, was in command of at least 17 different gesture-signs, at the age of 63 months, of 47 signs; and at the age of 73 months, of 51 signs (Fouts and Fouts and Van Cantfort 1989). It is also worth noting that the intraspecies communication by means of gesture-signs was not limited just to the interaction between mother-infant, but took place also among non-kin members of the group of chimpanzees (Fouts and Fouts and Schoenfeld 1984; Jaffe and Jensvold and Fouts 2002; Jensvold and Gardner 2000).

Another feature of the employment of gesture-signs discovered in the course of the AGSL project was that Washoe used the previously acquired gesture-signs to sign neither to humans nor to conspecific, but to herself, usually to comment on entities that were physically and/or and representationally (like pictures in journals) present (Fouts and Fouts 1993). Another variety of gesture-sign-use in which neither humans nor conspecifics participated were chimpanzees' employment of gesture-signs when encountering inanimate entities (Jensvold and Fouts 1993). So, for example, the male chimpanzee Dar signed "TICKLE" to a stuffed bear and then placed it between him and the fence; he always pressed his body against the fence when being tickled by humans. The same chimpanzee also put the stuffed bear into his pelvic pocket and signed to it "PEEKABOO"; a sign used when starting the game of hide and seek with humans. Washoe, for example, employed the gesture-sign "HAIRBRUSH" to refer to a tooth-brush, and then pretended to use it as a brush for her hairs.

The whole course of the AGSL project and its results led to a changed perception of the role, usability and efficiency of the method of S-R reinforcement. So, the Gardners declared, in a backward glance, the following (Gardner and Gardner 1989b, 16):

> At first, it seemed prudent to include as much of the technique of operant conditioning as might be compatible with the overall objective of raising an infant chimpanzee like a human child ... After all, a popular notion in 1966 was that human families must use variants of operant conditioning every day without realizing that they are doing so – hence in an inefficient, amateurish way.

But based on the course and the results of the AGSL project, their view changed markedly (1989b, 19)[5]:

> Operant conditioning was impractical as a method of teaching signs. Once a sign has been introduced into the vocabulary by whatever method, it was equally impractical to attempt to

---

5   See also (R. Gardner and Gardner 1998). For R. S. Fouts' comments, see (Fouts and Mellgren 1976) and (Fouts 1977).

reward all appropriate usage by prompt and consistent deliv-
ery of appropriate goods and services ... Meanwhile, all con-
nected discourse or conversation would certainly be disrupted
by such a procedure. For the objectives of cross-fostering, the
only practical way to proceed was to treat Washoe, Moja, Pili,
Tatu, and Dar as if they had an intrinsic motive to communicate with us, the way human parents treat human children.

In fact, what the Gardners proposed in the last quote is a
shift to methods employed in the practice of everyday communication among humans by means of language.

2.2 Chantek

Also participating in the experiments based on AGSL was
the male orangutan Chantek, under the guidance of L. W.
Miles. The conceptual basis she employed diverged from that
employed by the Gardners at the end of the 1960s and early
1970s.

Her basis involved the trio of semantic terms "sign"-
"reference"-"meaning" and the pragmatic term "use of
meaning of the sign", where a sign referred, when it is a
meaningful indicator, to the real world; this she spelt out
further into the following four points (1990, 511, 513, 524):

1) The sign can be used to designate an element in the real
   world, including objects and events that are not actually
   present.
2) It is not totally context dependent.
3) To the sign is assigned a realm of meaning that is cultur-
   ally understood in a shared manner.
4) This meaning is conveyed intentionally by the use of the
   sign;

But even when Miles employed quite a sophisticated semantic
terminology, she evaded the ultimate step of declaring that
semantics and pragmatics should take center stage in the

research into the language capabilities of great apes. Instead, she declared: "It should be noted that the goal of this research was not to demonstrate whether or not Chantek had acquired 'language' ... In contrast, the focus of Project Chantek is on a developmental perspective that seeks to identify the cognitive and communicative processes that might underlie language development". And as the reason for evading semantics and pragmatics, she stated "Linguists do not agree on the essential nature of language" (1990, 512).

Let me now turn to the results of the experiments with Chantek. His instructions started at the age of eleven months, and after one month he had already produced the first two gesture-signs, while in the second month he was combining gesture-signs into sequences. During seven years of experiment, his vocabulary expanded to 127 signs, meeting the acquisition criterion of spontaneous and appropriate usage in half of the days in a given month.[6] Those 127 signs fell into the following categories: objects, actions, foods, names of persons, animals, drinks, colors, locatives, other attributes (dirty, good, bad, hurt), emphasizers (more, hurry), personal pronouns (me, you), and places (Miles 1990, 517–518). Between ages 2 and 4.5 years Chantek was gradually able to refer to absent entities (the so-called *displaced reference*) as well as refer to new exemplars of a referent.

---

6   For more data on the dynamics of Chantek's vocabulary in this period and on the growth of the length of Chantek's signed utterances, see (Miles 1983).

3. The Yerkish Signs: Lana, Sherman, Austin, and Kanzi

3.1 Lana

The YL project was initiated by D. M. Rumbaugh and the first great ape participating in it was the female chimpanzee Lana. In this project, the media of communication were visuographic symbols, so-called *lexigrams*, designed from nine basic elements that could be combined into symbols that should stand for words. These lexigrams were embossed on the keys of a computer keyboard composed of several 5-by-5 key-matrices.[7] Table 2 lists these nine basic elements and some of their possible combinations into words.

Table 2 *Nine basic symbols of the Yerkish language and some of their possible combinations into words*

---

7 For technical detail of YL, see E. von Glasersfeld's (1974), (Glasersfeld *et al.* 1973), and (Rumbaugh *et al.* 1973).

By selecting and depressing particular keys on which lexigrams were embossed, Lana could actuate a vending device, controlled by the computer, to provide her with a particular food (e.g., M&M, a piece of banana, a sweet potato).

Lana was initiated into language not by teaching her from the outset particular lexigrams words but by teaching her whole "stock sentences". Initially, the lexigrams of a stock sentence, say, "PLEASE MACHINE GIVE M&M.", were interconnected in such a way that the depression of just one lexigram lighted all the keys and arranged them into the right order, which led to a dispensing of the respective food. The whole stock sentence was then gradually split up. First, the lexigrams for the requested entity were separated and Lana had to depress two keys: "PLEASE MACHINE GIVE" and "M&M.". Then, she had to press three keys with the lexigrams "PLEASE MACHINE", "GIVE", and "M&M.". Next, she had to press four, "PLEASE", "MACHINE", "GIVE", and "M&M."; then, five lexigram keys had to be depressed, "PLEASE", "MACHINE", "GIVE", and "M&M" and the key for the period.[8] Next, the keys were randomly distributed among a larger number of keys, and finally, the location of the keys on the keyboard was regularly changed before a new experimental session began.

E. von Glasersfeld, as the person in charge of creating the lexicon, the grammar, and the rules of concatenation of lexigrams of YL, commented on the idea driving the creation of the lexicon delineating the universe of discourse possible in the framework of YL as being highly *anthropocentric* (1974, 18) in the sense that the *primatologists and behavior special-*

---

8   Here the lexigram "Please" signaled to the computer software an impeding request and the lexigram for the period, its termination.

ists involved in the YL project decided in advance which goals Lana could view as worth pursuing by means of communication in YL, and then the YL was set up accordingly (Glasersfeld 1977, 95).

Let me now turn to the method employed at the beginning of the YL project. Like the Gardners in the AGSL project, the preferred method initially intended by Rumbaugh and his collaborators was that of operant conditioning. Rumbaugh declared: "Initial training would be simple, fundamental operant conditioning whereby the subject, Lana, would come to master the basic discriminations and responses necessary to address the computerized system" (Rumbaugh and Warner and Glasersfeld 1977, 89). In fact, the very idea to employ a computerized system went back to this method (Gill and Rumbaugh 1977, 158):

> With regard to the intensity of training, it was decided that Lana would live in the language environment 24 hours a day. There, her linguistic expressions would provide repeated, reinforcing engagement with the system, since she would have to obtain all of her necessities and social interactions by making appropriate requests of it.

What were the results of linguistic experiments with Lana? She manifested skills of learning new lexigrams and the ability to hold to *syntactical* rules of stringing lexigrams into sentences. So, for example, when facing a set of three stock sentences – two syntactically correct and one incorrect – she was able to eliminate the syntactically[9] defective one with a success rate of 90 %. And she was able to complete stock sentences stopped by the human experimenter at any lexigrams.

---

9   Defective was either the ordering of lexigrams or a word was added in excess.

She also displayed the ability to apply previously learned expressions to novel situations.

Such an application could be seen in her mastery of color names.[10] She named color in a sentence – for example, "BALL WHICH-IS RED" – where the expression "WHICH-IS" was classified in the YL as an *attributive marker* (Glasersfeld 1977, 101). Based on this mastery she was then able to generate an expression, by means of a description, for an item she wanted but for which she lacked a separate lexigram name. This item was an orange held by the experimenter, Timothy Gill, outside Lana's room and the exchange that ensued between them was as follows (Rumbaugh and Gill 1977, 178–179):

Tim: "?WHAT COLOR OF THIS."
Lana: "COLOR OF THIS ORANGE."
Tim: "YES."

…

Lana: "?TIM GIVE WHICH-IS ORANGE."
Tim: "?WHAT WHICH-IS ORANGE."
Lana: "?TIM GIVE APPLE WHICH-IS GREEN." (At this point Lana frequently confused the lexigram keys for the green and orange colors.)
Tim: "NO APPLE WHICH-IS GREEN."
Lana: "?TIM GIVE APPLE WHICH-IS ORANGE."
Tim: "Yes." (And he gave it to her.)

Yet another of her language capabilities is worth describing here in full detail. Initially, Lana used lexigrams as names of items she could, once correctly placing the lexigrams in a pre-defined sequence of lexigram-order, receive as a reward – for example, the lexigram "M&M" into "PLEASE MACHINE GIVE M&M.". The question then was whether she would be able to name (that is, refer to) objects in another context – namely, once asked for the name of an item shown to her,

---

10 On this see (Essock and Gill and Rumbaugh 1977).

and where the crucial lexigram presented to her on the projector would be "NAME-OF".[11] In a sequence of tests, she demonstrated the capability to understand the meaning of this sign. When the experimenter put the question "?WHAT NAME-OF THIS." and before pressing the key with the lexigram "THIS" presented to Lana an item, say, a banana, Lana correctly responded by "BANANA NAME-OF THIS.".

In fact, her ability to employ the lexigram "NAME-OF" went well beyond communication encounters where a reward could be directly envisaged by her. In one situation, the lexigram "SLIDE" was placed on her keyboard and then the experimenter produced on the keyboard the sentence, "PLEASE MACHINE MAKE SLIDE.", at which point a slide projector was turned on by the computer and slides started to appear in front of Lana. Her spontaneous reaction to this was the production of the sentence "SLIDE NAME-OF THIS."

An even more innovative employment of the lexigram "NAME-OF" took place in the following circumstances.[12] Lana was taught the lexigrams "CAN" and "BOWL", referring to a metal can and a bowl, respectively. Next, Lana was subjected to the following experiment. The experimenter Timothy Gill approached her with a bowl, a metal can and a cardboard box, the latter being filled by him with M&Ms while being watched by Lana, who was not acquainted with the lexigram for box. Then, the following exchange ensued:

Lana: "?TIM GIVE LANA THIS CAN."
Tim: "YES." [And hands her the empty can, which she at once discards.]
Lana: "?TIM GIVE LANA THIS CAN."
Tim: "NO CAN." [He could not give her the can because he had given it to her already.]

---

11 On this see (Gill and Rumbaugh 1974).
12 On this see (Glasersfeld 1974) and (Rumbaugh *et al.* 1975).

Lana: "?TIM GIVE LANA THIS BOWL."
Tim: "YES." [And hands her the empty bowl.]
...
Lana: "?TIM GIVE LANA NAME-OF THIS."
Tim: "BOX NAME-OF THIS."
Lana: "YES."
Lana: "?TIM GIVE LANA THIS BOX."
Tim: "YES." [Whereupon Tim gives the box to her and she extracts the M&Ms.]

Noteworthy in this exchange is that unlike the Lana's employment of the lexigram NAME-OF in a declaration being prompted by a request for a name of an item, now Lana by herself actively asked for a name by relating the lexigram "GIVE", *not to a physical item* such as M&Ms, as she had been trained to do, but to a *name* of a physical item. Such a departure from a trained routine on her part came as big surprise; Glasersfeld commented on this, "A spontaneous generalization of GIVE, not foreseen by the grammar, since NAME-OF has not been classified as a possible object of GIVE!" (1974, 53–54). I will explain in Part 6 the reason why this generalization came as a surprise to the team of the YL project.

Another language experiment worth mentioning pertains to an experiment in which Lana's reaction to Gill's deliberate lie in communication encountered was tested. The experiment started with Gill's placement of cabbage into the vending machine instead of monkey chow, which Lana preferred much more than cabbage. The exchange that then ensued between them was as follows:[13]

Lana: "?YOU PUT CHOW INTO MACHINE."
Tim: "CHOW IN MACHINE."

---

13 This is my transcript of the video *Amazing Apes* from minute 2.45 to minute 3.54. This video can be accessed at https://www.youtube.com/watch?v=HiWDKXRzSmU.

[This exchange took place four additional times. Lana then came back with]
Lana: "?CHOW IN MACHINE."
Tim: "YES."
Lana: "NO CHOW IN MACHINE."
Tim: "?WHAT IN MACHINE."
Lana: [No response]
Tim: "?WHAT IN MACHINE."
Lana: "CABBAGE IN MACHINE."
Tim: "YES CABBAGE IN MACHINE."
Lana: "?YOU MOVE CABBAGE OUT-OF MACHINE."
Tim: "YES" [and proceeded to remove the cabbage]

Two utterances produced by Lana are notable: "NO CHOW IN MACHINE." and "CABBAGE IN MACHINE.". In both of them Lana referred to a state of affairs in the extra-linguistic world. In Part 6, I will provide a semantic analysis of the types of reference given in these utterances as well as a pragmatic analysis of the exchange between Tim and Lana listed above.

3.2 Sherman, Austin, and Kanzi

The YL project underwent several important changes starting in 1978, when E. S. Savage-Rumbaugh became involved and the results of language experiments with Lana were reevaluated. This reevaluation was based on an analysis of both the training procedures to which Lana had been subjected and the behavior she manifested when subjected to language experiments. So as she was subjected to operant conditioning, the question arose: "Was Lana's use of various sentences to obtain specific items she desired simply a complex set of instrumentally conditioned responses? ... Were Lana's 'stock sentences' simply conditioned responses?" (Rumbaugh and Savage-Rumbaugh 1978, 124, 125). Savage-Rumbaugh's answer to this was a qualified *no* (1978, 125), because the results of experiments with Lana showed that she was able to employ lexigrams independently of context.

But at the same time Savage-Rumbaugh also gave a *negative* answer to the question of whether Lana displayed a mastery of a language. An experiment in which the reward for a successfully accomplished procedure would not be the named item but another item, wherein Lana's performance after initial success rapidly deteriorated, provided evidence for the assertion that Lana had not displayed a mastery of language. According to Savage-Rumbaugh, the reason for this decline in performance was that Lana experienced as reinforcement something other than the named item (Rumbaugh and Savage-Rumbaugh 1978, 125).

Savage-Rumbaugh's interpretation of the YL experiments with Lana led the researcher thus to a conclusion in the form of an antinomy: no instrumental conditioning, thus, mastery of language, and no mastery of language, thus, instrumental conditioning. In order to escape it, she coined the term "levels wordness," which she explained as follows (Rumbaugh and Savage-Rumbaugh 1978, 125):

> [T]here is the strong indication that only gradually does the lexigram accrue symbolic representational value ... Eventually ... word lexigrams do serve as symbols to represent things and events not necessarily present or ongoing in the here and now. It is likely that symbolism results from training experiences that decontextualize a word lexigram ... Rather than saying that a chimpanzee "knows" or "has" a certain number of words, it appears that it is more appropriate to speak of "levels of wordness." "Words" may, at first, be simply conditioned associations between symbol and referent. Later, important but generalized correspondences between word usage and environmental occurrences become perceived.

But there still exists a higher level of wordness where the entity experienced and/or physically handled is different from that which is referred to in a linguistic exchange; Lana was not able to reach this level (Rumbaugh and Savage-Rumbaugh 1978, 126). Savage-Rumbaugh's conclusion was, "We can find no definite demonstration that Washoe ... [and] Lana ... used symbols representationally" (Savage-Rumbaugh and Rumbaugh and Boysen 1980, 55).

In the penultimate quotation given above appears the verb "to know". Savage-Rumbaugh applied this verb to great apes who became acquainted either with AGSL, YL, or PT when she asked: "Does the ape inherently *know* that a sign, a lexigram, or a plastic chip can stand for an object that may be absent in time and space? Does the ape inherently know that a *name* may be used to convey information to another animate about that object?" (Savage-Rumbaugh and Rumbaugh and Boysen 1980, 50) – that is, "do they know that symbols can reference objects, people, places, events, action, states, relationships, and so forth?" (Savage-Rumbaugh 1981, 35).

Her questions pertain to those great apes who underwent some training in the employment of a language, so, seemingly, the best way to answer them would be *to put them directly to these great apes*. But can this question at all be put to these apes in the respective language into which they were introduced by human experimenters? In Part 6, I will give an answer to this question.

Based on these critical reflections, Savage-Rumbaugh moved to experiments in which the chimpanzees Austin and Sherman participated. These experiments, in her view, were more advanced than those that had been conducted with Lana in at least the following two aspects: First, the named entities would differ from those which Sherman and Austin received as rewards; she referred to this as the difference between *labeling* and *requesting* (1981, 38). Second, the experiments should test the great apes' ability to establish a relation

31

not only between a lexigram and its referent, but also among the very lexigrams, when their physical referents were absent.

In one of these experiments (Savage-Rumbaugh *et al.* 1980; Savage-Rumbaugh 1981), during the training phase, Sherman and Austin were, first, trained to label by means of lexigrams placed on the keys of a keyboard three inedible items (stick, key, money) and three edible items (orange, bread, beancake). Next, the chimpanzees were taught to sort these six items into two categories, food and tool, via the "edible versus inedible" functional distinction, by placing them into two separate bins: one for foods and one for tools.

Once this task was mastered, a separate lexigram "TOOL" for tool in general and lexigram "FOOD" for food in general were introduced on the keyboard in addition to the bins with their particular objects. Once the correspondence between objects-bin and produced lexigram was mastered by the apes, the bins were removed and the apes had to label each of the six objects either with the lexigram "FOOD" or the lexigram "TOOL" (i.e., they had to create pairs like, for example, <"FOOD", stick>) by pressing the corresponding lexigram on the keyboard.

After this level of training was successfully mastered, photographs of the six objects mentioned previously were introduced, initially by taping them to the objects themselves, while the apes were asked to assign via the keyboard lexigram "FOOD" or "TOOL" to the respective pairs, for example, <stick, photograph of the stick>. Once this task was mastered, the objects were removed and the apes had to label only the photographs with one of the two available keyboard lexigrams "FOOD" or "TOOL". After this task was mastered, the last and crucial phase of the experiment was initiated.

The task was to assign via the keyboard one of the two available lexigrams, "FOOD" or "TOOL", to the lexigram-image, placed into a plastic casing, for each of the six objects. Initially, the tool-lexigram or the food-lexigram had to be assigned to a photograph-lexigram pair and then, after this was mastered, the photograph was removed. Finally, new items (5 food-items and 5 tool-items) were introduced. Figure 1 reproduces the results obtained by Sherman and Austin together with an iconic representation of the sequences of training and testing these apes underwent.[14]

Fig. 1 Flow chart of experiment with items learned in the left column and items tested in the right one

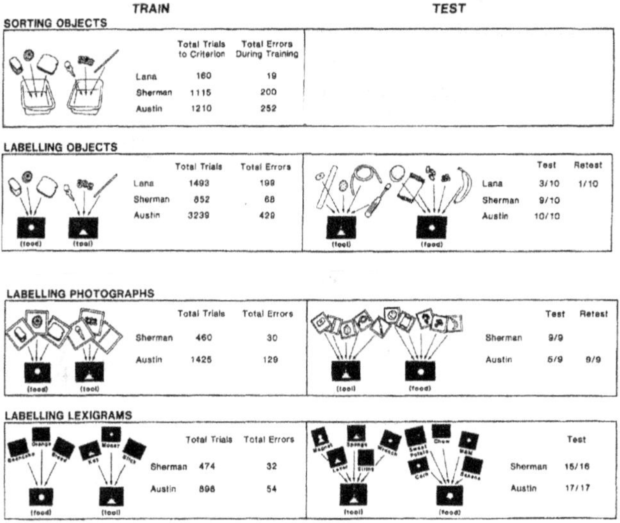

TRAIN / **TEST**

SORTING OBJECTS

	Total Trials to Criterion	Total Errors During Training
Lana	160	19
Sherman	1115	200
Austin	1210	252

LABELLING OBJECTS

	Total Trials	Total Errors			Test	Retest
Lana	1493	199		Lana	3/10	1/10
Sherman	852	68		Sherman	9/10	
Austin	3239	429		Austin	10/10	

LABELLING PHOTOGRAPHS

	Total Trials	Total Errors			Test	Retest
Sherman	460	30		Sherman	9/9	
Austin	1425	129		Austin	5/9	9/9

LABELLING LEXIGRAMS

	Total Trials	Total Errors			Test
Sherman	474	32		Sherman	15/16
Austin	898	54		Austin	17/17

---

14 The left column reproduces the training sequences, the right, the testing sequences. So as Lana failed in this experiment, I eliminated the results for her.

The fact that both of the apes were able to assign a produced lexigram to a lexigram-image in a plastic casing, by pressing a key on the keyboard, was interpreted by Savage-Rumbaugh as their ability to "use one symbol to classify another, thereby forcing them cognitively refer to the specific referent of one symbol and ... assign to it a class of functionally related items" (1981, 51) and as evidence that they "could group the symbols alone into classes of food and tools" (1984a, 297) and that "they could group, not just objects, but also lexigrams into proper superordinate categories even when the referents were absent" (1987, 289).

From this interpretation it is, however, not clear what Austin's and Sherman's performance really were. It was, for certain, not a classification of the shape of the lexigram-images in plastic casings by means of the shape of the lexigram produced on the keyboard. Were they able to express the relation between the referents of the lexigrams – for example, between the referent of "ORANGE" and "FOOD" and/or between "WRENCH" and "TOOL"? If yes, then we face the questions: *What is referent of the lexigram "FOOD"?* and *What is referent of the lexigram "TOOL"?*

What one can say here is that the referent of "FOOD" is different from the referent of "M&M", of "BANANA", and so forth. Similarly, the referent of "TOOL" is different from the referent of "WRENCH", of "STRING", and so forth. In order to give a positive answer to those two questions, one has to enter the territory of logical semantics; I will do so in Part 6.

In the course of the YL project Savage-Rumbaugh performed a shift in the reflections on methods of research into the language capabilities of great apes. As early as in (Rumbaugh and Savage-Rumbaugh 1978), these researchers put *cognition and generalization linked to language learning* into mutual opposition with *reinforcement and operant condi-*

tioning of behavior linked to language learning, while assigning priority to the former.

However, contrary to the views presented in this paper, in a later article Savage-Rumbaugh claimed that "the behavior-analytic framework, and the procedures devised to produce language-skills in apes, provide strong support for several of the major positions set forth in Skinner's (1957) *Verbal Behavior*" (1984b, 223).

The critical reflection on operant conditioning continued in her critique of the means available to *Science*, where the latter is understood as the sum of methods employed in primate behavioral research. The critique runs as follows (Savage-Rumbaugh 1999, 161–162):

> The real difficulty here is that living organisms do normally interact, and the "observer" stance is not the same as the "participant" stance. We cannot treat primates like particles of matter, for which the mixing and treatment procedures of one chemist can be replicated by those of another. Primates have memory, and the history of one's past interactions determines the nature of future interactions. One participant is not equal to another because their histories are not equal. And an observer is not equal to a participant because an observer stands outside. ... In the arena of ape language, the participants are also the researchers. There are no other "observers" standing by observing the participants.

The final and explicit statement about the departure from the method of operant conditioning in the YL project came in the form of an explicit declaration about a shift towards the methods used in "participant-based ethnographic studies" (Savage-Rumbaugh *et al.* 2005, 311) that rely "upon insight, intuition, and analysis of the observers, who are ... participant observers in the classical anthropological tradition. Narrative accounts, by definition, describe events. They do not predicate events, nor do they focus upon quantitative data" (Savage-Rumbaugh and Fields 2006, 223).

The final break from operant conditioning in the YL project was concurrent with experiments with the bonobo Kanzi. The specificity of these experiments was an outcome of unforeseen consequences of the classical lexigram training to which Kanzi's mother had been subjected (Savage-Rumbaugh *et al.* 1986). During her training, Kanzi accompanied his mother and was separated from her only after reaching the age of 30 months. Then, unexpectedly, Kanzi started to use lexigrams from a keyboard. From this, Savage-Rumbaugh concluded that Kanzi had acquired the ability to use lexigrams just by observing and imitating and not through operant conditioning.

This led to a profound change in the way Kanzi and the human experimenters around him interacted. Instead of the explicit training regimen to which Lana had been subjected, Kanzi was integrated into a way of life wherein he moved around in the woods surrounding the laboratory compound. Here he could pick up objects he initially indicated by means of lexigrams and that were placed in certain locations, and at the same time he was interacting with the experimenters by means of a mobile lexigram keyboard. Owing to this new way of interaction with humans, his vocabulary increased from initial 8 lexigrams to 356 lexigrams, while Lana's vocabulary never exceeded 140 lexigrams.[15]

---

15 For a quantification of this increase see (Savage-Rumbaugh and Rumbaugh and McDonald 1985, 659–663).

4. The Plastic Token Signs: Sarah

The PT project was initiated by D. Premack and its first experimental subject was the female chimpanzee Sarah. In the project's early phase Premack viewed as its goal the search for an answer to the questions: "Can apes be taught language? Although this question is of biological import, it may ultimately be more important to the fundamental question, What is language?" (Premack 1971a, 808).

In order to answer these two questions, Premack set up two lists, one stating the functions (exemplars) an animal should be able to do in order to provide evidence of language, and another stating the training procedures (recipes) that would enable the production of the respective functions in the animal. The first list's items concern certain aspects of what Premack spelled out in (1971a): (i) *words*; (ii) *sentences*; (iii) *questions*; (iv) *metalinguistics* (using language to teach language); (v) *class concepts* (for example, color, shape, and size); vi) *copulas*; (vii) *quantifiers* (all, none, several); and (viii) *the logical connective if-then*. All elements (i) through (viii) were mastered by Sarah.[16]

The physical objects used in Sarah's instruction were plastic tokens of different colors, sizes, textures, and shapes, with a metallic back that could be stuck to a magnetic board. Tokens had the status of words; a sequence of tokens combined to form a sentence was placed on the board vertically. Figure 2 shows this arrangement (Premack 1971a, 809).

---

16 On this see (Premack 1970b), (Premack 1971b) and (Premack 1971c).

Fig. 2 *Individual plastic tokens stand for words, while tokens combined in a vertical sequence form a sentence*

Let me now turn to Premack's words, sentences, metalinguistics, and class concepts. He viewed words as essential units of language and he introduced them in the course of feeding routines. First, Sarah was several times given a piece of food – say, a banana – from a tray placed between Sarah and the experimenter. Next, she was presented with the pair

<token for "BANANA", banana>, the token was located close to her, while the banana was beyond her reach. Once she placed the token expressing the word "BANANA" on the board, she was given the banana. The training continued with the experimenters simultaneously changing both elements of the pair <plastic token, food>. Sarah very quickly became proficient in the skill of assigning a plastic token to a fruit.

Next, Sarah was tested for her ability to assign the right token to the presented fruit, where the test was based on a variation of the elements in the pair <plastic token, food>. Initially, she was presented with two would-be words and just one fruit, so that she was tested for her ability to assign the right word to the presented fruit. Next, her preference for the very fruits was determined, and then she was presented with a set of words. Once her choice of a fruit matched her choice of a particular word, then this was viewed as a test of her naming capability. By means of this test it was possible to rule out that in the test <token-word$_1$, token-word$_2$, fruit presented> she would choose the "wrong" token-word, simply because she did not want the fruit presented to her, but the fruit absent, and thus chose the "wrong" token-word.

Finally, the test was varied to include the name of a donor of a fruit; Sarah had to place the donor token and the fruit token on the board in a specific order – say "MARY BANANA" not "BANANA MARY", "RANDY BANANA" not "BANANA RANDY". The number of words in word chains was gradually increased from two to three and eventually to four words. Figure 3 gives examples of these chains (Premack 1970b, 55).

Fig. 3 Chain of words with a predetermined order taught to Sarah

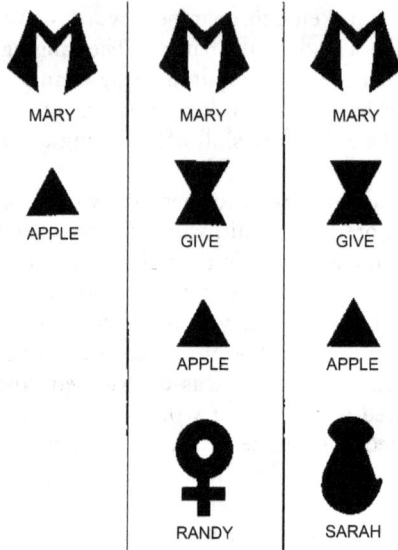

After Sarah mastered the production and understanding of chains of words, she was instructed to produce and understand sentences. Premack understood the production of sentences as displaying, in contradistinction to a simple chain of words, an internal organization (1970a, 113); he taught this organization to Sarah in two forms: one, he called, "two-term relation" and another, he called, "hierarchical organization."

I deal here with the case of two-term relations. Here the taught words were "ON", "FRONT OF" and "SIDE OF", which were combined with names "RED", "GREEN", "BLUE", and "YELLOW", which Sarah acquired earlier as names of cards differing just in color but indistinguishable otherwise. The experiment consisted of three steps. First, a red card was placed on a table in front of Sarah and the experimenter put on the board three plastic tokens making up the sentence "GREEN ON RED". Then Sarah was handed a green card and was induced to place it on the red card. Next, a green card was placed on the table, Sarah was presented with the sentence "RED ON GREEN" and was handed the red card and then induced to place it on the green card. Finally, she was handed both cards at once and successively was presented with the sentences "GREEN ON RED" and "RED ON GREEN"; for both she was able to produce the right arrangement of cards.

In the second step, she was presented with sentences involving all four color words and had to arrange the cards in the order corresponding to the order described in the respective sentence. She was also able to master this task. While in the first and second steps she had to produce an order of cards corresponding to the order expressed by a sentence, in the third step the task was reversed. The experimenter produced an ordering of color cards and Sarah had to produce on the board the corresponding sentence using the three words handed to her in the form of plastic tokens. She was also able to manage this task. She also successfully mastered the similar three-step experiments with the words "FRONT OF" and "SIDE OF".

Sarah, like Lana in the YL project, also successfully managed the employment of the metalinguistic expression "NAME OF". Figure 4 shows that the words "APPLE" and "BANANA" and the words "NAME OF" and "NOT-NAME OF" are paired with the named object (Premack 1970b, 58).

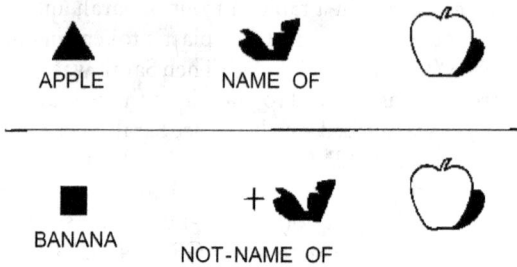

Fig. 4 Relations of naming and not naming between plastic tokens and the named object

Let me now turn to the two groups of experiments conducted by Premack with Sarah. The initial experiment (1971b) with Sarah was based on her previous mastery of the expressions "BANANA" and "APPLE", which Premack called *"names of objects"*, and on her mastery of the interrogative marker "?". She was presented with the questions "RED ? APPLE" and "YELLOW ? BANANA" and the only word she could use was "COLOR (OF)". She substituted "COLOR (OF)" for "?" and obtained the sentences "RED COLOR (OF) APPLE" and "YELLOW COLOR (OF) BANANA".

Next, she was taught negative instances by being faced with the questions "RED ? BANANA" and "YELLOW ? APPLE" while having at her disposal only "NOT-COLOR (OF)". This enabled her to form the sentences "RED NOT-COLOR (OF) BANANA" and "YELLOW NOT-COLOR (OF) APPLE". This procedure was followed by the simultaneous presentation of "COLOR (OF)" and "NOT-COLOR (OF)" to the expression "RED ? BANANA" and "YELLOW ? APPLE". Once Sarah was able to form the expressions "RED NOT COLOR (OF) BANANA" and

"YELLOW NOT COLOR (OF) APPLE", the final step consisted of a test of her ability to apply the expressions "COLOR (OF)" and "NOT-COLOR (OF)" to new names, like, for example "CHERRY". She was able to pass this final test of transference by creating the sentence "RED COLOR (OF) CHERRY"

The next type of experiment was based on her additional mastery of the expressions "COLOR", and "SHAPE", which Premack labeled as *"property names"*. The procedure used here was the same as in the previous type of experiment. This experiment showed that Sarah was able to form sentences, for example, "RED IS COLOR", "ROUND IS SHAPE", "RED IS-NOT SHAPE" and "ROUND IS-NOT COLOR", and via transference, "YELLOW IS-NOT SHAPE".

Once Premack viewed expressions like "BANANA", "APPLE" and "CHERRY" as *names of objects* and expressions like "COLOR" and "SHAPE" as *names of properties*, he faced the following problem which he described as follows (I have inserted language expressions in square brackets as examples) (1971c, 70):

> In teaching Sarah a word to denote class membership, we did not make adequate allowance for the abstract character of the relation. What is meant in saying that class membership is an abstract relation? Compare, for example, color as the relation between red and apple ["RED COLOR (OF) APPLE"] with class membership as the relation between apple and fruit ["APPLE IS FRUIT"] (or red and color ["RED IS COLOR"]). In the case of color ["COLOR (OF)"], it is possible to provide referents for both terms in the relation ["RED" and "APPLE"], but in the case of class membership ["IS"] one of the terms ["FRUIT"] is elusive. In saying, X member of Y [for example "APPLE IS FRUIT"], X ["APPLE"] is easy to provide a referent for, but Y ["FRUIT"] is not. That is, the

43

class member is easily instanced, but the class itself is not. (Should we treat a list of class members as a referent for the class, and thus introduce class membership as the answer to the question. What is the relation between, say, "apple" and "apple, banana, orange, and raisin"?)

This shows that he encountered *explicitly* the problem that was, as shown earlier, *implicitly* given in Savage-Rumbaugh's experiments with Austin and Kanzi, who were able to assign the keyboard lexigram "FOOD" to the lexigram-images in plastic casing "BANANA", "AP-PLE", and so forth, and the keyboard lexigram "TOOL" to the lexigram-images in plastic casing "WRENCH", "STRING", and so on.

The fact that Premack's problem reappeared in the YL project indicates that it has at least some importance. As I will show in Part 6, the key to its solution will be a treatment of the expressions "COLOR (OF)" from "RED COLOR (OF) APPLE" and "IS FRUIT" from "APPLE IS FRUIT" in the framework of one and the same semantic approach.

Premack's approach, in my view at least, lacked such a unified treatment. This can be seen from Table 3 which lists the vocabulary he taught to Sarah (1970b, 56).

Table 3 Vocabulary taught to chimpanzee Sarah

SARAH'S VOCABULARY

I. Names
Sarah (Chimp)
Mary
Ann
Randy
Debbie
John
Jim
Gussie (Chimp)

II. Verbs
prefer
open
close
give
take
insert
cut
stir
cook
eat
not eat
drink
wipe
put/place
is
is +plural = are
is not
is +pl+not = are not
dry
smoke
peel

III. Colors
yellow
blue
green
red
orange
brown

IV. Foods
Fruit
  orange
  apple
  banana
  peach
  pear
  fig
  apricot
  grape
  date

Starch
  cracker
  bread
  cookie
  sugar

Candy
  chocolate
  caramel
  Cracker Jack

peanut
peanut butter
honey
jam
milk

V. Objects
key
table
dress
shoe
blouse
cup
spoon
can
crayon
paintbrush
paper
bubbles
flashlight
toy top
sponge
bottle
broom
garbage can
dust pan
basin
dish
pail (2 words for)
soap
toothbrush
toothpaste
comb

VI. Concepts, Adjectives, Adverbs
name of
not-name of
size of
not-size of
shape of
not-shape of
color of
not-color of
equal (same)
not-equal (different)
if, then
big
little
round (2 words for)
triangle
square
good
bad
one
several
all
none
much
question mark
yes
no, not
none
now
on
to the side of
in front of

45

This vocabulary is based on a strange blend of expressions unified on the basis of a *syntactic* classification (adjectives and adverbs) with expressions unified by means of a *non-syntactic* classification – the so-called concepts – however, without a clarification of what a concept should stand for. In addition, this blend was employed side-by-side with a division of language expressions naming real-world entities as belonging to three classes (for color, for foods and for objects, the latter meant, probably, as naming non-food entities); again, the criterion used in this threefold classification of expressions was not explained.

5. Mathematics in Great Apes

Because my focus is on *language* experiments with great apes, mathematical experiments with them can be divided into three groups: experiments with no numerals, experiments that include numerals and physically present items, and experiments with numerals only.

The first were experiments where no *numerals* – that is, no *signs of language of mathematics* – were involved but they still aimed at testing certain mathematical capabilities in great apes. Examples of these are the experiments described in (Rumbaugh and Savage-Rumbaugh and Hegel 1987) and (Rumbaugh and Savage-Rumbaugh and Pate 1988). These experiments showed that great apes were capable of selecting from two trays presented to them, one containing the larger amount of food items, with their choice being based on the operation of addition of pairs of items on each of the two trays.

The second were experiments in which *numerals were involved alongside with physically present items* and in which the aim of the experiments was to test great apes for their ability to label the numerosity of these items by means of numerals. So, for example, in (Matsuzawa 1985) it is shown that the female chimpanzee Ai displayed the capability to pair numerals embossed on keys on a keyboard with the number of elements from sets displayed on a computer screen and where the sets contained up to six elements. A similar type of experiments was described in (Boysen 1993), in which several apes were able to associate numerals placed on cards with the corresponding number of food items on a dish. In this way they acquired mastery of the numerals "1", "2" and "3"; additional numerals "0", "4", "5", "6", "7" and "8" were introduced directly without that association.

The third were experiments based on the exclusive use of numerals – that is, *mathematical language was* (if you will) *their only medium*. From the point of view of my focus on language, I classify the third type as the most advanced type of mathematical experiments with great apes. As far as my knowledge goes, they were for the first time presented in (Boysen and Berntson 1989). Here the female chimpanzee Sheba was described as performing the operation of addition. This operation was performed on numbers ranging from 0 through 4 and the pairs on which it was performed by Sheba were (1, 0), (1, 1), (1,2) (1, 3), (2, 0), (2, 1) (2, 2), and (3, 0).

6. What is Language?

6.1 Language

The works of the experimenters involved in language experiments with great apes abound with hints as to the need to move to a semantic treatment of the notion of language and to the issues such a treatment should address. I mention here three such hints.

First, Premack noted that the language experiments with great apes faced the question, "What is language? On the basis of what properties do we identify a system as a language? ... Linguists ... did not address questions of this general kind" (Premack 1986, 4). And as to the discipline that should find answers to those questions he declared that "the animal language inquiry had more to gain from epistemology and philosophy of language than from linguistics. Reference, meaning, and truth are philosophical predicates, not linguistic. These predicates have the abstractness we are looking for" (1986, 8).

Second, from Premack's conceptual triple *reference-meaning-truth* as the proper subject-matter of the discipline, he labeled "philosophy of language," the first (reference) became the subject matter of special interest both for the AGSL and the YL projects in their later stages. So Savage-Rumbaugh declared the following (1981, 35):

> The issue of referential capacity is absolutely fundamental to any concept of language. It is more basic than the issue of modality and more important than the issue of syntax. If the symbols of a language are not referential, if they cannot be used to represent things, people, places, events, states, and relationships that are absent in space and time, then they have no more information value than the emotive facial or

gestural expressions. To put it bluntly, communication that is not referential is not language, regardless of what other qualities it may possess.

In a similar manner, the Gardners declared: "Although concern with grammar has occupied so much the efforts of developmental psycholinguistics, in our view it would be a mistake to neglect method and theory in the study of reference ... unless verbal behavior refers to objects and events in the external world, it cannot communicate information" (R. Gardner and Gardner 1989a, 230).

Third, in addition to the notion of reference, one of the projects (the AGSL project), encountered during its undertaking the issue of the *ideality of the shape of gesture-signs*. For this project, to be successful it was necessary to instruct the great apes in such a way that they would acquire the ideal patterns of signs. On the basis of these signs, the apes should then produce, by employing parts of their bodies, the physical realizations of these ideal patterns; I will label the latter "*ideal signs*", while the physical realization of these ideal signs I will call "physical signs"[17].

According to Miles, the orangutan Chantek acquired these ideal signs because he was able "to respond to his caregiver's request that he improved the articulation of a sign. When his articulation became careless, a caregiver would ask him to SIGN BETTER. He then would sign in a slow, emphatic way, and with one hand, he would put the other hand in the proper sign shape ... Chantek understood that the sign had to approximate an *ideal shape*" (1990, 530; emphasis added).

The YL and the PT also faced the issue of ideal signs and their physical realization by the apes – however, not in such

---

17 My terminology here is based on (Cmorej 1985) and (Cmorej 2001, 13–18).

an explicit way as the AGSL project did. The reason for this was that great apes involved in the YL and PT projects had to choose the right physical realization from a set of physical object – namely, a lexigram embossed on the keys of a keyboard or a set of plastic tokens – *given to them in a ready-made form.*

In contradistinction to this, the great apes in the AGSL project had to *produce* the physical realizations of the ideal signs by employing parts of their own bodies. In a gesture-sign language, the physical realization of an ideal gesture-sign has to match at least the following: a certain shape of hand or hands, a certain location of the hand or hands in space, the movement of the hand or hands, and a facial expression.[18] So, a fully-fledged notion of language has to integrate at least the following four notions: *ideal sign, physical realization of the ideal sign, referent,* and *meaning.*[19]

Let me start with physical signs. They stand for entities having a certain space-time existence determined by the way in which they were produced. Thus, for example, we have acoustic physical signs produced in the vocal tract, graphic signs produced by handwriting or by print, and kinesthetic gesture-signs produced by certain parts of the body. Each of these groups has as its basis certain *ideal signs*, which are not physically realized space-time entities, but abstract, sensually not perceptible entities that have respect to their physical realizations a prescriptive, normative status in the sense that they delineate how the physically produced spoken or written/printed or gestured sign should look.

---

18 For a classical treatment of the first three criteria, see the works of W. C. Stokoe, Jr. (1960; 1972).

19 My terminology here is based on P. Cmorej's (1985; 1998; 2001, 13–18).

The process of the physical realization of ideal signs then determines the type of *language modality* being produced. For example, the physical realization of ideal signs of an *acoustic* language determines that a *spoken* language (*speech*) is being produced, while by the realization of ideal gesture-signs one obtains a gesture-sign language.

It is also the case that each type of physical realization corresponds to a certain type of *physical reception* of the physically realized sign. The vocally realized ideal acoustic sign requires an acoustic reception; the physical realization of both ideal written/printed and ideal gesture-signs requires a visual reception, while the physical realization of a dactyl sign requires a reception via touch. The unity of realization and reception of the set of physically realized ideal signs of a particular language is the basis for its reproduction inside a community actually using it and for the reproduction of that language across generations.

This means that for two or more subjects to engage in communication by means of a language, they have to share not only a set of ideal signs, but also the same corporeal constitution – the latter in order to be able, in a physically identical manner, both to produce and to perceive the physical realization of ideal signs.

The differentiation between ideal signs of a language from their physical realization enables us to overcome a widely spread misconception about gesture-sign languages – namely, there is just one gesture-sign language used by all the deaf. The truth is, however, that while all the deaf employ the same medium for the physical realization of ideal gesture-signs and all the deaf employ the visual mode for reception of this realization, still these gesture-sign languages differ by their respective ideal gesture-signs and thus, in turn, also by their respective physical realizations.[20]

Let me now turn to the notion of *meaning*, say, as an example, of an ideal English sign-chain realized in written English as "president of the United States of America".

Even if this ideal sign-chain and its realization are different from the French ideal sign and its realization in written French as "président des États-Unis d'Amérique", each of them is still tied by means of a special type of relation, which I label the relation of *denotation*, and that holds between the respective ideal sign and its meaning – that is, the sign *denotes* the meaning as its *denotatum*. This relation is established by the semantics of the respective language; in order to know this meaning it suffices to know this semantics. An important consequence of the existence of the relation of denotation is that: what is standardly understood to be the *meaning of an expression* stands in fact for the pair <ideal sign, denotatum>.

If we look again at the example, another feature of the relation of denotation comes to the surface. Here I refer to the fact that this relation is of a purely *semantic* nature and completely independent of the actual situation in the extra-linguistic world; in my example the actual holder of the office labeled in English "president of the United States of America". This extra-linguistic entity I label *referent*, while the relation between an ideal sign and its actual referent in the extra-linguistic world I label *referring*.

Thus, in order to determine the actual referent of an ideal sign, it is necessary to turn to the actual situation in the extra-linguistic world; this turn takes its course via the mediation of

---

20 By "all deaf" I mean here all the deaf with the exception of the deaf-blind.

the ideal sign's denotatum. The relation of referring between an ideal sign and its actual referent in the extra-linguistic world thus stands for the triple <ideal sign, denotatum, actual referent in extra-linguistic world>. My explication of the notion of language by means of its constitutive elements can be expressed as shown in Figure 5.

Fig. 5 Relations among the constitutive elements of a language

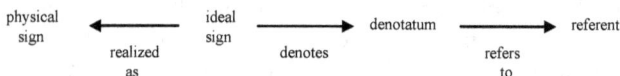

| physical sign | ← realized as | ideal sign | → denotes | denotatum | → refers to | referent |

What are the implications of the notion of language understood in such a way for the very language experiments with great apes? First, it can explain why, as mentioned earlier, the method of operant conditioning based on the stimulus-response (S-R) theory employed in the early phase of the AGSL and the YL projects lost its efficiency once the great apes acquired the mastery of employing at least some language expressions in the communication with experimenters. The reason for this is that once the ape mastered the pairs of the type <ideal sign, denotatum>, then the chain stimulus-response was interrupted by being mediated by such pairs; and once the linguistic exchange between a great ape and a human experimenter gradually extended in time, the mediating link between these extremes got longer and longer so that the great ape had to focus more and more on longer and longer chains of pairs <ideal sign, denotatum> as well as on the intertwining of production and reception of physical realizations of ideal signs. Under such pressure of ever-increasing chains, the reward status of the physical stimulus for the very great ape retreated into the background.

Second, the intervention of the chains of the type <ideal sign, denotatum> strikes back on the physical entities at the extremes in the sense that these extremes become meaningful for the subjects involved in the linguistic exchange. This can be seen, for example, in the addition experiment with Sheba. For her, the cards, which she physically perceived and manipulated, acquired a meaning – namely, they embodied numbers. Therefore, once the physically given extremes became meaningful for the ape who acquired a language, he or she did not perceive them anymore as just the physical entities to be given to him or her as a food reward. This explains, at least in my view, the success of Savage-Rumbaugh's separation of Austin's and Sherman's *naming* (*labeling*) of an item from their *requesting* of the item as a reward.

The relation of the constituent elements of a language as expressed in Figure 5 can be applied for the characterization of several types of language expressions; my focus here is on declarative sentences as well as on subsentential language expressions: names of properties, names of relations, and proper names. Here I draw on the results of modern intensional semantics, in which the term "intension" is used instead of "denotatum" and the term "extension" is used instead of "referent"; I will hold to this terminology.[21]

The intension of a *declarative sentence* stands for a *proposition* that assigns to the sentence its *actual truth-value* as its extension in the extra-linguistic world. The proposition of a declarative sentence is the prime carrier of its truth-value in the sense that it determines the truth conditions of the sentence expressing this proposition. Thus, what is standardly viewed as the understanding of the meaning of a declarative sentence stands for a grasping of the proposition expressed by it and thus for an understanding of what makes (or would make) this sentence true.

---

21 For an introduction to intensional semantics, see (Carnap 1956).

From the point of view of the theory of speech acts, focusing on the employment of language expression in *communicative acts*, declarative sentences acquire in such acts the form of *assertives*.[22] These are uttered with a claim to truth and the communicative counterpart to the utterer accepts or refuses this claim to truth, that is, takes the position of "yes" (agreement) or "no" (disagreement) with regard to this claim.[23] *The presence of declarative sentences in communicative acts – that is, when communication becomes propositionally differentiated – indicates, in my view, that communication is already accomplished in the medium of language.*

Let me now turn to what logical semantics views as *names of properties* and *names of relations*. Examples of these names I take from the declarative sentences "APPLE WHICH-IS ORANGE", "BANANA IS FOOD", and "WRENCH IS TOOL" from the YL project, and from the declarative sentences, "RED IS COLOR" and "RED COLOR (OF) APPLE" from the PT project. By eliminating the expression "BANANA" from the sentence "BANANA IS FOOD", one obtains the expression "_____ IS FOOD", where "_____" indicates an open position which can be filled by a language expression. The expression "_____ IS FOOD" is viewed in modern semantics as the name of *property* – that is, its intension (denotation) is a property.

What is its extension (referent)? In order to find this, let me turn to the expression "BANANA" in the sentence "BANANA IS FOOD". So as the former ("BANANA")

---

22 On the basics of theory of speech acts see Searle's (1969; 1975; 2001). My employment of the term "assertive" is based on Searle's classification of illocutionary acts.

23 Here I profited from Habermas' (1983).

was used by Lana, Austin and Sherman to refer not just to one and only one individual entity but to several belonging to a set labeled by that expression, each and all members of this set can be referred to by the expression "IS A BANANA". In a similar manner it holds that the extension of "IS (TO BE) FOOD" is a set whose elements are sets of individual bananas, sets of apples, etc. Thus, we can say that the intension of "TO BE A BANANA" specifies a set that is a subset of the set specified by the intension of "TO BE FOOD".

By the same analysis we can demonstrate that "TO BE WRENCH," "TO BE RED," "IS (TO BE) TOOL," and "IS (TO BE) COLOR" are names of properties, that is, of intensions that specify the respective set as extension.

Another semantics is given in the expression "APPLE WHICH-IS ORANGE" appearing in the exchange described earlier between T. V. Gill and Lana. So, as here "APPLE" refers to an individual entity shown to Lana – that is, it is in fact a proper name – the expression's "_____ WHICH-IS ORANGE" denotation (intension) is a property.

Let me now turn to the sentence "RED COLOR (OF) APPLE" produced by Sarah. By eliminating in it both "RED" and "APPLE", one obtains "_____ COLOR (OF) _ _ _ _", where "_____" and "_ _ _ _" indicate open positions that can be filled by language expressions.

The expression "_____ COLOR (OF) _ _ _ _" is viewed in modern semantics as a name of *relation*, that is, its intension (denotation) is a relation while its extension is a set of ordered pairs. In the PT experiments with Sarah this set involved the ordered pairs <yellow, banana>, <red, apple> and <red, cherry>.

A more complicated type of a name of relation can be found in the previously reproduced encounter between Tim and Lana. By removing from her question "?YOU PUT

CHOW IN MACHINE." the triple "YOU", "CHOW", "MACHINE", and from her request "?YOU MOVE CAB-BAGE OUT OF MACHINE." the triple "YOU", "CAB-BAGE" and "MACHINE", one obtains expressions "_____ PUT _ _ _ _ IN . . . ." and "_____ MOVE _ _ _ _ OUT . . . .", respectively. Both expressions are viewed by semantics as names of relation whose extensions are sets of ordered tri-ples; here these triples are <Tim, chow, machine> and <Tim, cabbage, machine>. The same semantics is given in the sen-tence "MARY GIVE APPLE SARAH" in Figure 3, which was taught to Sarah in the PT project. Here the expression "_____ GIVE _ _ _ _ TO . . . ." is a name of relation whose extension is the ordered triple <Mary, apple, Sarah>.

The semantics of a name of a relation whose extension is a set of ordered pairs has to be distinguished from the semantics of the expression "_____ NAME-OF _ _ _ _", whose employment was mastered both by the chimpanzee Lana in the AGSL project and the chimpanzee Sarah in the PT project.

While it is possible to obtain from the sentence produced by Lana, correctly written as "'BANANA' NAME-OF THIS", by eliminating both "'BANANA'" and "THIS", the expression "_____ NAME-OF _ _ _ _", which carries two open position, is different in terms of semantics from that of "_____ COLOR (OF) _ _ _ _", which also carries two open positions.

The difference between the two expressions is made clear by examining the semantics of the expressions, which can be substituted into the open positions in "_____ NAME-OF _ _ _ _" and in "_____ COLOR (OF) _ _ _ _", respectively. In the former, the language expression "'BANANA'" refers to the *language expression* 'BANANA', while the language expression "THIS" refers to an *extra-linguistic* entity. Con-trary to this Janus-faced feature of "NAME-OF", in "_____

COLOR (OF) _ _ _ _", the expressions which can be substituted for both open positions, for example "RED" and "APPLE", "YELLOW" and "BANANA", and so forth, *refer to extra-linguistic entities.*

The Janus-faced nature of the expression "NAME-OF" brings to the surface the specific reflexivity of languages employed in communication, namely: some of its expressions do not refer to extra-linguistic entities but to expressions of this very language. This reflexivity can be employed by language users for the purpose of clarifying the semantics of certain expressions from this very language to their communicative counterparts.[24]

Finally, let me deal with the semantics of *proper names.*[25] This semantics is different from the semantics of both the names of properties and names of relations. Proper names have a simple semantics because they refer always to the same referent under all circumstances. In order to have such semantics, they have to be introduced into the respective language. This introduction stands for an act of naming, for example by *ostension (pointing)* to the physically present referent, which, from the point of view of semantics, stands for an introduction of the pair <ideal sign, referent> into the respective language. Once this introduction is performed, the proper name can function in the respective community using this language where it spreads among its members.

Earlier, when dealing with the use of declarative sentences in communicative acts, I referred to the speech-act-theory's view, according to which declarative sentences in such acts acquire the status of assertives. This theory presents in fact a

---

24 Here I draw on (Habermas 2001, 73).
25 Here I found M. Zouhar's (2002a; 2002b; 2003) very helpful.

broader typology of communicative acts. So as I will employ this typology for an evaluation of the language capabilities of great apes from the respective language projects, I present here the following typology of these acts:[26]

1. *Assertives* (*representatives*); they commit the speaker or gesture-signer to the truth of the expressed proposition.
2. *Directives*; they stand for the attempt of the speaker or gesture-signer to get the communicative counterpart to do something by means of, for example, orders, commands, or requests.
3. *Comissives*; they commit the speaker or gesture-signer to a future course of action, for example, by promising.
4. *Expressives*; by means of them the speaker or gesture-signer expresses his or her feelings and attitudes; for example by means of apologies, thanks or congratulations.
5. *Declarations*; by means of them the speaker or gesture-signer brings about an institutional or conventional state of affairs by declaring that it obtains, for example, by pronouncing a couple as being married.

From the point of view of the employment of language in communicative acts, it is easy to see that the explication of the notion of language by means of its constitutive elements, as represented in Figure 5, is incomplete. The latter focuses on the referential function of language while neglecting its communicative function. Figure 6 takes into account both functions.[27]

---

26 Here I used Searle's typology of illocutionary acts presented in (1975; 2001).
27 On the intertwining of these two functions, see (Habermas 2003, 2–7).

*Fig. 6 Intertwining of the communicative and referential functions
        of language*

*Fig. 6 Intertwining of the communicative and referential functions
        of language*

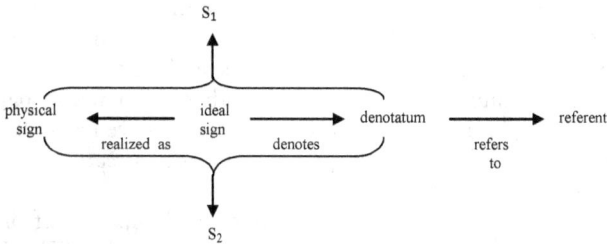

The braces indicate that for the two subjects, S_1 and S_2 to engage in communication by means of a language they have to share a set of ideal signs with their respective denotata and the same corporeal constitution – the latter in order to be able in a physically identical manner both to produce and perceive the physical realization of these ideal signs.

6.2 AGSL project revisited

Regarding the semantic approach to language expression and the classification of speech acts, the following results of the AGSL project should be mentioned.[28] First, as shown in Table 1, Washoe was able to produce declarative sentences embedded both in *assertives* and *directives*, the latter in the form of orders and requests.[29] Second, Washoe employed the expression, "DIRTY", not only to name feces and stains on clothes, shoes, etc., but transferred that term and combined

---

28 I obtained some of the information reported here through email communication with R. A. Gardner.

29 For other produced declaratives – for example, by the chimpanzee Tatu – see (Gardner and Gardner 1980, 345).

it with other terms in order to express her displeasure, say, with a person who did not comply with her request.[30] The same combination was also detected in the gesture-signs of orangutan Chantek (Miles 1990, 518). I place such combinations into the category of *expressive* speech acts.

The communicative acts of the other chimpanzees participating in the AGSL project can also be viewed as falling into one of the three categories *assertives*, *directives*, and *expressives*.[31]

Third, as to subsentential expression, chimpanzees from the AGSL project produced declarative sentences like "THAT HOT" and "SANDWICH HOT" (Gardner and Gardner and Nichols 1989, 71). Here the expression "_____ HOT" can be viewed as a *name of a property*.

Fourth, chimpanzees in the AGSL project produced the directive "YOU TICKLE ME"; the very expression "_____ TICKLE _ _ _ _" embedded into this directive can be viewed as a *name of relation*.

Washoe as well as other chimpanzees in the AGSL were able to refer to themselves by means of their respective *proper names*. As to Washoe, her first encounter with her own name was when signed by an experimenter who *pointed* to Washoe's image in a mirror while gesture-signing to her the question "WHO'S THAT?" In a similar manner, the orangutan's proper name "CHANTEK" was introduced by L. W. Miles in the course of experiments where this orangutan faced his own image in a mirror (Miles 1994).

---

30 On this, see (Fouts and Fouts 1993).
31 I used the data from (Leeds and Jensvold 2013), which lists, in addition to Washoe's communicative acts, also those of the chimpanzees Tatu, Moja, Dar and Loulis. Expressives produced by Chantek are listed in (Miles 1990, 518).

The metalinguistic expression "NAME-OF" was not used in the experiments performed by the Gardners because they employed only gesture-signs found in vocabularies used by humans and they presupposed that the gesture-sign for "NAME-OF" was not given in those vocabularies.

6.3 YL project revisited

As shown previously in one of the dialogue of Lana with T. Gill, Lana was able to produce two declarative sentences: "NO CHOW IN MACHINE." and "CABBAGE IN MACHINE.", to refer to the state of affairs in the vending machine.[32] In addition, again, as shown earlier, Lana, Austin, and Sherman were able to produce both names of properties and names of relations. As to the proper names employed in the YL project, the names "LANA", "SHERMAN" and "AUSTIN" were not introduced by pointing combined with phrases like, for example" "YOU LANA", because they were not viewed as helpful in the acquisition of these proper names.[33]

Instead, these proper names were introduced by means of sentences such as "TIM TICKLE LANA", "LANA TICKLE TIM", "SHERMAN GROOM SUE", and so on. These sentences were used to indicate what the agent (Tim; Sue) was going to do with the recipient or, alternatively, to ask the chimpanzee to be the agent who would act on the recipient. So, the chimpanzees' names were introduced in the context of

---

32 This ability should be taken into account in the dispute concerning whether great apes are capable of producing declarative sentences. Regarding this dispute, see (Rivas 2005), (Lyn and Russell and Hopkins 2010), and (Lyn *et al.* 2011).

33 This information as well as those mentioned below, I obtained by email communication with E. S. Savage-Rumbaugh.

the physical presence of named individuals and not by means of other language expressions – for example, by means of descriptions of these individuals.

From the previous transcripts of communications with Lana, it is obvious that she accomplished communicative acts of the *assertive* as well as of the *directive* types. As to the *expressive* communicative acts, involving expressions such as "BAD", "HAPPY" and "GOOD", they were not used with Lana, Austin and Sherman but were used with Kanzi as well as several other bonobos involved in the YL project. Let me now turn to the following three features of the YL project.

First, as mentioned previously, Savage-Rumbaugh employed the expression "to know" in order to ask questions like, "Does the ape inherently *know* that a sign, a lexigram, or a plastic chip can stand for an object that may be absent in time and space?" (Savage-Rumbaugh and Rumbaugh and Boysen 1980, 50), and "[D]o they know that symbols can reference objects, people, places, events, action, states, relationships, and so forth?" (Savage-Rumbaugh 1981, 35).

Worth mentioning here is that by employing the expression "to know" in such questions, Savage-Rumbaugh asked something much stronger than just "Do the apes *use* a sign, a lexigram, or a plastic chip in order to refer to objects, events, action, relationship, and so forth?" Rather, she was asking about the *epistemic status that that use has for these very great apes*. The simplest way to answer those questions would be, seemingly, to address these questions directly to these great apes – for example by means of a question addressed to those subjects who had already mastered the YL.

This question, however, could not be put to a great ape participating in the YL experiments simply because the lexicon of YL did not contain the lexigram "KNOW-THAT", which, from the point of view of modern logic, stands for an epistemic operator. Here surfaces a specific feature of the

YL – namely, the criteria which were at work in its creation. These criteria were, using Glasersfeld's terminology, *anthropocentric* by their nature in the sense that YL *was constructed based on views what a great ape is linguistically capable of*.

With respect to that epistemic operator, it meant that the lexigram "KNOW-THAT" was not integrated into the lexicon of YL because its *creators presupposed that Lana would not be able to employ it in communication*. Therefore, given the structure of the YL, Lana could neither be asked in an experiment the question that would involve the expression "?LANA KNOW-THAT _____", nor could she produce a declarative sentence that would involve that operator.

We encounter here a specific feature of the language experiments with great apes – namely, *the language capabilities great apes manifest in these experiments and, thus, their results, are codetermined by the structure and richness of the particular language they were taught by humans*. For YL, for AGSL, and for the PT language thus holds what Habermas stated with respect to any language – that "the structures of language determine the channels of possible interaction" (Habermas 1988, 74).[34]

I use here the term "codetermine" in order to express the fact that as in the case of communication among humans (Figure 6), the corporeal constitution of great apes, in addition to the structure of language, enters as a factor determining the communicative exchanges with humans, and thus also the outcome of experiments that target their language capabilities. For the language experiments with great apes to

---

34 This claim of Habermas goes back to his reinterpretation of Wittgenstein's "*The limits of my language* mean the limits of my world" (*Tractatus* 5.6) into "the boundaries of action are drawn by the boundaries of language" (1988, 72).

get off the ground, the employed types of ideal signs and their physical realizations have to fit that corporeal constitution.

Second, worth noting is that some of the views on the language capabilities of great apes which were at the basis of the creation of the YL, in fact, *underestimated* great apes' linguistic capabilities. This became readily seen in the experiment testing Lana's capability to employ the lexigram "NAME-OF". The initial presupposition reflected in the construction of the YL was that, once an experimenter presented to Lana the question "?WHAT NAME-OF THIS.", while showing her a food item, say, a banana, Lana would produce on the keyboard the sentence "BANANA NAME-OF THIS." in order to be rewarded with this item. What the underestimation of her language capabilities consisted of was the presupposition that while Lana could react to the question involving that semantic indicator by using it in a declarative sentence, *she could never ask for the name of an item by employing that operator in a question in which it would be combined with the lexigram "GIVE". A great ape, it was presupposed, could ask to be given just a food item*, say by means of "PLEASE MACHINE GIVE M&M.". Lana, however, contrary to this constraint imposed on her by means of YL taught to her, was able to escape it by stating the question, "?TIM GIVE LANA NAME-OF THIS."

In Part 7, I will present a view on which anthropocentrically based criteria the construction of a language taught to great apes should be based and whose application would at least partially avoid the possibility of an underestimation of these great apes' linguistic capabilities.

Third, and finally, the use of the language by great apes involved in the YL project displayed – compared to its use by great apes involved in the AGSL project – the specific feature that they neither used it for spontaneous communication with their conspecifics nor in the instruction of their offspring.[35]

6.4 PT project revisited

As shown above, Sarah was able to employ both names of properties and names of relations in the context of declarative sentences – that is, to employ them communicatively in the context of assertives. She was also able, as shown in Figure 3, to understand the proper name "SARAH", used to refer to her, and to employ it productively. However, I was not able to find out from the works of D. Premack whether Sarah was taught this proper name by means of direct reference or by another method. Neither was I able to find out if she was able to employ communicative acts of the directive and expressive types.

Let me now turn to the problem encountered by Premack that I mentioned above in Part 5. This problem was related to sentences of the form "X MEMBER OF Y" – for example, "APPLE IS FRUIT" – and, to be more specific, to the lack of making an adequate allowance for the abstract relation being at work. In Premack's view, while one can easily find a referent of the expression "X" (e.g., for "APPLE"), one cannot find a referent for "Y" which he understands as a name of a class; that is, while "the class member is easily instanced ... the class itself is not" (1971c, 70).

The problem Premack faced has its roots in a wrong semantical classification of the relation expressed by "X MEM-

---

35 The experiments described in (Savage-Rumbaugh and Rumbaugh and Boysen 1978a) and (Savage-Rumbaugh and Rumbaugh and Boysen 1978b) were artificially created encounters between Austin and Sherman taking place in a laboratory where the YL was employed by them.

BER OF Y" as class membership. The problem disappears when, for example, both "IS (TO BE AN) APPLE" and "IS (TO BE A) FRUIT" are viewed as names *denoting* properties and *referring* via the latter to sets. This denotation can be viewed as a semantic allowance for the abstract relation, which was missing in Premack's approach.

This view on the denotata and referents of both "IS (TO BE AN) APPLE" and "IS (TO BE A) FRUIT" in turn enables one to find an answer to Premack's question, "Should we treat a list of class members as a referent for the class, and thus introduce class membership as the answer to the question, What is the relation between, say, 'apple' and 'apple, banana, orange, and raisin'?" (1971c, 70). While the set of individual apples is the referent of "IS (TO BE AN) APPLE", the referent of "IS (TO BE A) FRUIT" is the set involving a set of individual apples, a set of individual oranges, etc. This means, using the terminology of intensional semantics, the denotation (intension) of "IS (TO BE AN) APPLE" specifies, as its extension, a set that is a subset of the set specified by the intension of the expression "IS (TO BE A) FRUIT".

6.5 The addition experiment revisited

In order to apply the semantic approach to language from 6.1 to the addition experiment with Sheba, it is necessary to understand the latter's semantics. It is a lack of awareness of this semantics which is symptomatic for the description of the addition experiment with Sheba in the Boysen-Berntson article. This can already be seen from its abstract, in which the following appears (Boysen and Berntson 1989, 23):

> A chimpanzee *(Pan troglodytes),* trained to count foods and objects by using Arabic numbers, demonstrated the ability to sum arrays of 0–4 food items placed in 2 of 3 possible sites. To address representational use of numbers, we next baited

> sites with Arabic numbers as stimuli ... the animal was able to
> select the correct arithmetic sum for arrays of foods or Arabic
> numbers under novel test conditions. These findings demon-
> strate that counting strategies and the representational use of
> numbers lie within the cognitive domain of the chimpanzee.

The employment of the expression "Arabic number" indicates that the authors do not differentiate between, on the
one hand, *numerals* as *signs* and, on the other hand, *numbers*. The former refer to the latter; that is, *Arabic numerals
(signs)*, not numbers, refer to numbers in a similar way as,
say, Roman numerals (signs) do. This confusion between the
naming entity and the named entity in this paper goes well
beyond its abstract as can be seen from expressions appearing
in its main body like, for example, "black Arabic numbers"
(1989, 24).

Based on the above given differentiation between an ideal
sign and its physical realization, it follows that what is described in the abstract in (Boysen and Berntson 1989) as
"baited sites with Arabic numbers as stimuli" stands in fact
for sites baited with the physical realizations of ideal Arabic
numerals (ideal signs). This then means that what the addition experiment with Sheba in fact proved was, she was
able to master that part of the language of mathematics that
involved triples of the type <ideal numeral/sign, physical realization of the ideal sign/numeral, number>.

What has been said up to now about the addition experiment with Sheba shows that the semantic treatment of
language as presented in Part 6.1 can also be extended to the
language of mathematics. However, there exists an important
semantic difference between, on the one hand, the declarative
sentences like – for example, "TO BE BANANA IS (TO BE)
FOOD", and, on the other hand, the declarative sentences
of mathematics – for example, "$2 + 1 = 3$". While the truth
of the first depends on the existence of the extensions (refer-

ents) of "TO BE BANANA" and of "IS (TO BE) FOOD" in the extra-linguistic world, the truth of "2 + 1 = 3" does not.

What this means is that while in the former one has to differentiate, from the point of view of semantics, between the intension (denotatum) and the extension (reference) of the expressions "TO BE BANANA" and "IS (TO BE) FOOD", this difference does not make sense for Arabic numerals "2", "1" and "3" from the declarative sentence "2 + 1 = 3".

This would, from the point of view of Figure 5, seemingly lead to an impoverished semantics for expression from the language of mathematics. But for the language of mathematics, in addition to the entities listed in that figure, it is convenient to introduce yet another semantic entity, namely, *construction*.[36] What a construction stands for can be explained by turning to the set of pairs of numbers listed in the above description of the experiment with Sheba.

The operation of addition performed by Sheba on certain pairs of numbers always yielded the same number: for the pairs (1, 1) and (2, 0), it led to the number 2; for (1, 2), (2, 1) and (3, 0), it led to the number 3; and for the (1, 3) and (2, 2), it led to the number 4. We thus have two different constructions that yield the same number 2; three different constructions that yield number 3; and two different constructions that yield the number 4. If we view each of this construction with regard to its respective result as a way of givenness of a number and the number as an intension (= extension), then we can say that Sheba was able to perform several mutually different constructions yielding one and the same intension (= extension).

---

36 For an introduction of this entity into the semantic treatment of language of mathematics see (Tichý 1995).

Following (Tichý 1995), the notions of numeral, number and construction can be expressed by means of Figure 7.

Fig. 7 The semantics of numerals as signs denoting numbers

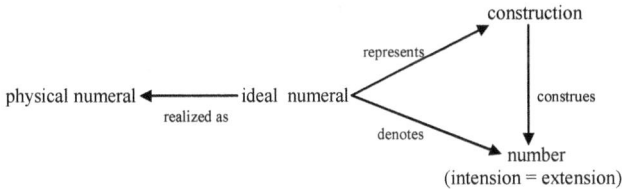

Sheba's ability to perform constructions drew of course on a certain *base*, from which natural numbers could be integrated into the construction and from which she could choose the corresponding result of the construction. In the experiment in which she participated, the base at her disposal was quite narrow as it involved just five elements (0, 1, 2, 3, 4).

Another important limitation of the addition experiment with Sheba was that while she explicitly faced signs for numbers, she *never explicitly faced* a sign for the operation of addition. It thus remains an open question whether she was at all aware that she was performing this operation. Of course, given the fact that she was tested only for the ability to perform the operation of addition, the introduction of the sign for this operation could be regarded *at least from the point of view of human experimenters* as being redundant.

It will not be redundant, once she will be tested for the ability to differentiate, not only among numbers, as she was in the addition experiment, but also among mathematical operations. Based on the application of the above given elements of semantics for the language of mathematics, I will propose in Part 9 new experiments aimed at testing great apes' ability to master the language of mathematics.

7. Methods of Research into Language Capabilities of Great Apes

I will approach the methods employed in language experiments with great apes from two points of view. First, from the point of view of a general method indicating that these experiments involve a combination of the method of understanding in the sense of understanding the meaning of communicative acts with the method of pursuing non-communicative strategic research aims. The second point of view will allow me to solve the problem of choosing anthropocentric criteria for the construction of a language taught to great apes.

The *general method* employed in all three language projects stands for the human experimenters' *and* great apes' grasping of the meaning of utterances produced in their mutual communicative encounters and, where the human experimenter's grasping is unified with his or her focusing, with respect to his communicative counterpart, on a strategic – that is, a noncommunicative – aim.[37]

What these two aspects of the general method stand for in detail can be understood if we turn to the encounter described earlier in which Tim (Gill) put a cabbage into the vending machine and uttered the declarative sentence "CHOW IN MACHINE". On the one hand, an exchange of communicative acts took place between Tim and Lana. Here the issue at stake faced by both was to understand what the communicative counterpart stated. This required from both an understanding of the intensions (denotata) of the expression used by the counterpart and in this case declara-

---

37 This differentiation between communicative and strategic aims goes back to (Habermas 1984).

tive sentences, which as assertives commit their utterer to the truth of these sentences, as well as communicative acts of the directive type.

On the other hand, and at the same time, the experimenter pursued, by purportedly uttering the factually false sentence "CHOW IN MACHINE", the aim of testing Lana's capability, first, to understand the claim to truth by the assertion of this sentence; second, to compare this claim with the actual state of affairs in the machine; and, third, to reject that claim to truth by stating "NO CHOW IN MACHINE".

This testing was, on the part of the experimenter, not only a noncommunicative aim superimposed on the communicative exchange between the experimenter and Lana. This superimposing was hidden to Lana as was the purpose of the experimenter to give a statement of a factually false sentence by which he wanted to deceive Lana with the aim of finding out whether she would be able to detect this deception. This of course required that the experimenter be continually engaged in linguistic exchanges with Lana.

Let me now turn to a delineation of a method that would enable us to escape the anthropocentricity of the creation of a language taught to great apes in the sense that it would not be based on the views of their human creator on the great ape's language capabilities.

In order to provide such reconstruction, I have to address one aspect of the understanding of the notion of language as given in Part 6.1 – namely, the relation of language to the extra-linguistic world.[38] As I have shown, the path to the extra-linguistic reference is channeled through the realm of intensions (denotata); only the intensions (denotata) enable beings who mastered a language to refer, in linguistic

---

38 Here I draw on (Habermas 2003, 17–22).

exchanges, to this world. Thus, from this point of view, language has, with respect to both these beings and the extra-linguistic world, an *extra-worldly (transcendental) status*.

But this point of view has to be modified with respect to the following factual information. Starting from Piaget, we have gradually realized that the command of a language by a human individual has also a beginning in time. This shows that language has not only an outerworldly but also an *innerworldly* status. From this then follows that what was presented in 6.1 as a notion of language seemingly constructed a priori was the description of an innerworldly given entity that is, from the point of a human individual, the result of ontogenesis.

For the sake of simplicity, I label the individual who is already in full command of a language due to his or her ontogenesis an *adult*. Therefore, the characteristics of language as well as the types of communicative acts described in Part 6.1, I ascribe to what I label a *human adult*.[39]

The clarification of the two statuses of language enables us to propose a new form of anthropocentrism, which should be at the basis of the creation of a language in which great apes should be instructed. Here I mean that this language's structure should be identical with the structure of language used by human adults.

The instruction of great apes in such a human language would then enable us to evaluate the degree to which these great apes can master a human language. Here, two situations could be obtained. Either the results of language experiments would show that great apes are capable of mastering the full structure of a language mastered by humans or, alternatively, they would show that great apes can master just a subset of this structure.

---

39 This ascription is performed also in (Dore 1975, 30).

The results of language experiments with great apes conducted so far indicate that great apes are able to master just a subset of this language structure. So, for example, while the great apes involved in the AGSL project were able to master at most 200 gesture-signs, that is much less than a deaf human infant, who can master at the age of 3–4 years more than 1,000 gesture-signs, both receptively and productively (expressively) (Andrews and Logan and Phelan 2008).

With respect to those two possible outcomes of language experiments, a paradox becomes apparent: as human adults instruct great apes in a language which at most is identical to *their own* language, *so the language experiments with great apes can never show that their language capabilities exceed those of human adults*. In order to show this, the experiments organized by *human adults* would have to employ a language with a structure *more complex than a language employed by these adults*.

My proposal of a truly anthropocentric approach to the construction of a language in which great apes would be instructed has, in fact, two precedents in the actual practice of language experiments with great apes. Here I refer to the approach applied by the Gardners in the language experiments based on the AGSL and by L. W. Miles in the classification of speech acts performed by Chantek, where this classification draws on a typology designed for the speech acts performed by human infants and where this typology, in turn, draws on J. R. Searle's typology of speech acts – that is, speech acts performed by human adults.

According to the Gardners, the success of the AGSL was based on the discovery and employment of technique based on the question: Which aspects of human language a great ape can acquire? (Gardners and Gardners 1980, 333):

Those who study the acquisition of language by human children have not found and do not expect to find a litmus paper test that can reveal just when this or that child has acquired its native language. What they do find is a pattern of development extending over years ... Gradually and piecemeal, but in orderly sequence, the language of the child evolves into the language spoken by the adult. This orderly sequence can be used as a yardstick to measure the achievements of nonhuman subjects ... Moreover, to the extent that the child and the chimpanzee exhibit similar patterns of development in the utterances that we can record, then a theory that accounts for the pattern found in the child must also be applicable to the pattern found in the chimpanzee.

From this justification of a methodological "yardstick", it is, however, obvious that in order to apply the yardstick in actual practice, one has to have at his or her disposal a description of what are the central characteristics of a language used by human adults. Only then one can view this description as a potential explanandum, which then has to be derived as an actual explanandum from certain initially prelinguistic-type, followed by protolingustic-type, and finally, language-type behaviors of a human child.

The description of the central characteristics of a language used by human adults, thus serves as the key in the search for the description of the relevant types of child behavior that should serve as the explanans for that potential explanandum and, by deriving it, turn it into an actual explanandum. This means that, once the *description* of the central characteristics of a language used by an adult is available, the description can be transformed into a *normative prescription* driving the search for the relevant data in the language behavior of a child. Once these data are unified into a description, they can serve as a key that drives, first, the search of the relevant data and, then, the creation of a description of the language mastered by great apes.

Based on such a clarification, the procedures described by Gardners can be represented by Figure 8.

Fig. 8 Procedures applied in the research into great apes' language capabilities

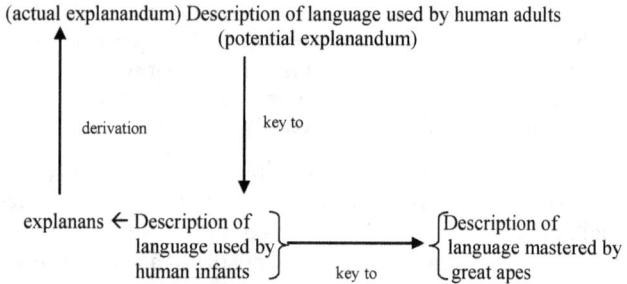

(actual explanandum) Description of language used by human adults
(potential explanandum)

derivation   key to

explanans ← Description of   Description of
            language used by ——————→   language mastered by
            human infants   great apes
                          key to

The lesson I draw from all this is that the *investigation into human adults' language on the basis of logical semantics and the typology of speech acts is a key to investigating human infant's language capabilities and, via the latter, a key to investigating language capabilities of great apes*. This I view as a justification of my choice of logical semantics and the classification of communicative acts provided by the theory of speech acts as the key to the analysis and reconstruction of the language experiments with great apes performed in the last 50 years.

The cycle going from the description of a language used by human adults as a potential explanandum to its rederivation, by the mediation of the description of a language used by human infants, as an actual explanandum belongs to the province of developmental psychology, while the application of the description of a language used by human infants as a key in the description of the use of a language used by great apes belongs to the province of comparative psychology. What comparative psychology should draw on is logical semantics and the classification of communicative acts by the theory of speech acts which, via the mediation of developmental psychology, enter the scene as a description of language used by human adults.

A unification of the procedure represented in Figure 8 by the downward pointing arrow, with the procedure represented by the vertical arrow in this figure was applied by L. W. Miles in the framework of the AGSL project for an investigation into the communicative acts produced by the male chimpanzee Ally.

This unification drew on the works of J. Dore (1974; 1975), where a cycle of derivations, represented in Figure 8 by the interconnection of both vertical arrows, was applied. Dore's conceptual starting point was Searle's division of illocutionary acts (1969, 16) into two parts: one which stood for a proposition and the other which stood for the illocutionary force of the utterance. Based on this approach, Dore provided an explanatory sequence with a successive derivation of descriptions of types of speech acts produced by human infants in their ontogenesis.

As the starting explanans, Dore employed a description of speech acts produced by a human infant, which he labeled as *primitive speech act* (PSA) delineated "as an utterance, consisting formally of a single word *or* a single prosodic pattern, with functions to convey the child's intention" (1974, 345) and that was produced by a child even before he or she produced whole sentences. Drawing on Searle's division, Dore represented the structure of a PSA as follows (1975, 34):

Fig. 9 The structure of a primitive speech acts according to
   J. Dore

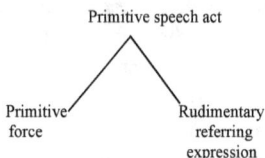

Dore described the transformation of the primitive type of speech act into a speech act already involving a rudimentary proposition by means of the following explanation; here the description of the primitive speech act stood for the explanans and the description of speech act for the explanandum (1975, 35):

Fig. 10 Explanatory derivation of the description of speech act
   from the description of a primitive speech act

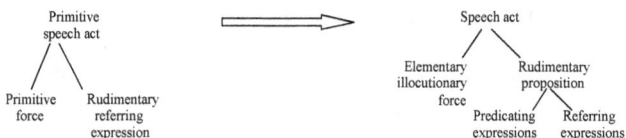

The next and final explanation given by Dore is as follows (1975, 35):

Fig. 11 Explanatory derivation of the description of speech act
with a more complete proposition from the description of
speech act with a rudimentary proposition

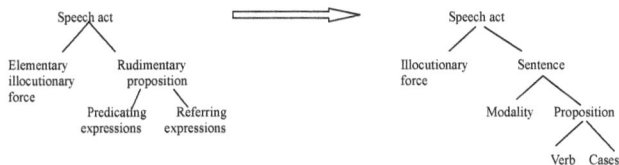

The starting point for Miles was Dore's list of speech acts falling within the category of primitive speech acts. This list involved nine distinct PSAs: labeling, repeating, answering, requesting (action), requesting (answer), calling, greeting, protesting, and practicing (Dore 1974, 346).[40] Miles applied this list as a conceptual key for the classification of speech acts performed by chimpanzees using AGSL. She explicitly linked the possibility to use this list as such a key to the *methodological assumption* that "the theoretical and performance criteria that can be ... applied to the use of language by deaf and hearing children can also be applied to the use of signs by chimpanzees" (1976, 593). The results of her experiments with the male chimpanzee Ally showed that from Dore's list, action requests and naming stood for 84 % of all his communicative acts.[41]

---

40 For details on these categories see (Dore 1974, 347) and (Dore 1975, 31).

41 For a detailed quantification and comparison with the results of experiments with human infants, see (Miles 1976, 594–596).

8. The Results of Language Experiments with Great Apes and the Turn to Social Science

The above given characterizations of results of language experiments with great apes were based on the employment of a semantic analysis of language expressions combined with a classification of illocutionary acts by means of the theory of speech acts. I summarize these results in Table 4.[42]

Table 4 Overview of performances of great apes in the respective language projects

Project / Performance	AGSL	YL	PT
name of proposition	+	+	+
name of property	+	+	+
name of relation$_2$	+	+	+
name of relation$_3$	?	+	+
proper name (*direct reference*)	+	+	?
semantic indicator "NAME-OF"	–	+	+
assertives	+	+	+

---

42 The sign "?" indicates that the results of the experiments reviewed by me do not indicate the respective performance. Here "name of relation$_2$" stands for a name of relation with two open positions, while "name of relation$_3$" stands for a name of relation with three open positions; "performance" refers to, with the exemption of the last line, production *and* reception.

Project / Performance	AGSL	YL	PT
directives	+	+	?
expressives	+	+	?
spontaneous transfer of signs among conspecifics	+	–	–

Based on such an overview of the results of language experiments with great apes performed in the last 50 years, it is possible to explain a peculiar feature of these experiments. While two of the three language projects – namely, the AGSL and the YL projects – were directly inspired by the S-R reinforcement theory, leading experimenters to use the method of operant conditioning, this theory and this method were in the course of these projects abandoned. In addition, the YL project was due to its results viewed by its authors as displaying the character of narrative ethnography.

So, while the AGSL and the YL projects initially approached the object domain of their investigation from a *natural-science* point of view, this point of view was gradually given up and the YL started to view the great apes from the point of a *social science*.[43] This change in the way how the object domain of the language experiments with great apes was accessed by the YL project can be justified by drawing on Habermas' reasoning with respect to the conceptual strategies a science-program, in order to be a veritable social science-program, has to apply when accessing its respective object domain (2001, 3–8).

---

43 For details on this see (Fields and Sagerdahl and Savage-Rumbaugh 2007).

According to Habermas, the crucial decision with respect to these strategies is to accept that linguistic meaning is constitutive for the object domain under investigation in the sense that this domain itself is determined by linguistic meaning. This decision, once made, according to Habermas, has several implications, two of which are of central importance for the abandonment of the natural science point of view of the great apes participating in language experiments and for YL project's claim that it acquired a social science point of view.

First, only if one accepts that linguistic meaning is constitutive for the object domain under investigation, can one differentiate between physical behavior and meaningful action, where this meaningfulness has its origin in the language being used in this domain. Second, only if one accepts that linguistic meaning is constitutive for the object domain under investigation, can one employ the method of understanding of that language meaning as the central method for accessing this object domain.

As I have described earlier, the experimenters involved in both the AGSL and the YL projects gave up the S-R reinforcement theory and the practice of employing the method of operant conditioning and started to interact communicatively by means of language with the great apes in a manner similar to the language interaction between human adults and human infants. In addition, as shown in Part 7, the method permanently employed by the experimenters in the process of interaction with great apes was that of understanding the meaning of language mediating this interaction.

While Habermas' reasoning goes *from* a metatheoretical decision a science program has to make in order to be a social science program *to* its implications for this program, I propose to reverse the order of reasoning – that is, those two aspects of the results of language experiments with great apes should be viewed as indicators that the very projects in the framework of which these results were achieved became part of social science.

9. New Mathematical Experiments: A Proposal

Based on the semantics for the language of mathematics presented in Part 6.5, I now propose several new experiments which, at least in my view, are worth to be performed with great apes in the future.

From the point of view of semantic notions of *construction*, *base*, and *number* in the sense of the *intension being construed*, the addition experiment with Sheba described in Part 5 can be viewed as follows. It involved three elements in the very construction that were pregiven (given to her prior to the conduct of the addition experiment): the operation of addition (+, for short) and two numbers (K1 and K2, for short); I express this construction as [+, K1, K2], while the intension (I, for short) is the construed entity. The addition experiment can then be expressed as [+, K1, K2] → I, where the arrow expresses Sheba's assignment of the intension to the construction.

The addition experiment performed by Sheba as well as the new mathematical experiments proposed by me are summarized in Table 5.

Table 5 Description of proposed new mathematical experiments with great apes

Pregiven Operation	[K1], [K2]	[K1], I	I	Ø
Addition [+]	I	[K2]	[K1], [K2]; [K3], [K4] ...	[K1], [K2] → I; I, [K1] → [K2]
Addition or Subtraction [+] or [−]	[+] or [−] → I	[+] or [−] → [K2]; [K2] → [+] or [−]	[+] or [−] → [K1],[K2]; [K1], [K2] → [+] or [−]	[+, K1, K2] → I; I, [+], [K1] → [K2] ... ...

The elements of the experiment that are pregiven to the great ape are shown under the heading "Pregiven". The left column lists the possible mathematical operation. In the first line the great ape faces the operation of addition as pregiven. The first cell in this line expresses that the great ape has to choose, with respect to pregiven signs of elements [K1] and [K2] from the construction, the sign for the corresponding number (intension) I; this task was mastered by Sheba.

In the next cell of this line the great ape faces, as pregiven, the sign for one element [K1] from the construction and the sign for the result of the construction, the intension I. The ape's task is to find the sign for component [K2] from the construction. In the following cell of the addition line, the ape faces, as pregiven, just the sign for the result of the construction – that is, the options from which he or she can choose are much less restricted as compared to the previous two cells. The task is now to choose a pair of signs for two elements [K1], [K2]; [K3], [K4]; ... from a construction that yields, together with the (pregiven) operation of addition, the (pregiven) intension I.

Finally, in the last column neither the signs for elements [K1], [K2] of the construction, nor the sign for the intension I, as the result of construction, are pregiven to the great ape. So, in the experiment represented by the last cell of the first line, the great ape is restricted only by the pregiven operation of addition and has thus several options. One is to choose a pair of signs for any two, [K1], [K2], and then find the sign for I, the other option is to pick at will signs for [K1] and I, and then find the sign for [K2], and so forth.

Of course, the background limitation of all the tasks represented in the first line is the base from which both the experimenter and the great ape can choose the natural numbers to be integrated into the construction as [K1], [K2], as well the intension I. Thus, the larger the base, the more unrestricted would be the experiments testing the capacity of a great ape for the operation of addition.

Another type of experiment to which great apes could be subjected would target their ability to master the operation of *subtraction*. Worth noting here is that S. T. Boysen subjected Sheba to the following experiment.[44] She presented to Sheba a tray with 5 bananas, which she then covered with a cardboard box with an open back. In full sight of Sarah, she next removed two bananas, while saying "BYE-BYE BANANA". Then she asked Sarah verbally to choose the card with the numeral corresponding to the number of bananas left on the tray. She was able to do this on the first trial.

This result was interpreted by Boysen as Sheba's ability to grasp the mathematics of the task. But this, at least in my view, need not hold. From the point of view of a human experimenter who is already in command of the operation of subtraction, the mathematics being here at work can be expressed by means of the declarative sentence "$5-2 = 3$". For Sheba, however, the task need not be at all related to the operation of subtraction. She could have perceived it simply as a task to assign a numeral to the number of bananas left on the tray.

What becomes apparent here is the limitation of mathematical experiments with great apes that do not take their course exclusively in the medium of the language of mathematics. In order to overcome such a limitation, a great ape should be taught to employ the sign "+" for the operation of addition and the sign "−" for the operation of subtraction.[45] The litmus test of a great ape's mastery of the respective operation would then be his or her ability to choose, when

---

44 This information was provided to me by S. T. Boysen.

45 In an email, S. T. Boysen informed me that she had no doubts as to the great apes' ability to learn the operator symbols.

facing a task, the correct sign of the operation. This type of experiment is described in the second line of Table 5.

Here the great ape could choose the operation sign of either addition or subtraction as an element of the construction ([+] or [−]). So, referring to the first cell of the second line, the ape, once facing as pregiven the signs for [K1], [K2], would have to choose a sign for one operation and then choose the sign for number/intension I. In the second cell of this line with pregiven signs for [K1] and I, the great ape could choose from two options: *either*, choose the sign for one operation and then choose the sign for [K2] *or* choose the sign for [K2] and then choose the sign for the operation. The tasks faced by a great ape described in the remaining third and fourth cell of this line should be evident from Table 5.

Another type of mathematical experiment, going beyond those described in Table 5 and introducing a new type of language expression could be performed by drawing on the introduction of the sign "=", once combined with the mastery of the task described in the third cell of the first line. Once the ape mastered both the employment of that sign and that task, the ape should be tested for his or her ability to place on one side of the symbol for identity an expression with the structure "[+, K1, K2]" and on the other another expression with the structure "[+, K3, K4]". This would thus test the ape for his or her ability to form declarative sentences with the structure "[+, K1, K2] = [+, K3, K4]", as given for example in "1 + 3 = 2 + 2". Once a great ape is capable of forming an expression with that structure, this could be viewed as a mastery of the semantics of declarative sentences stated in the language of mathematics as well as of the expression "=", that is, of what is expressed in English as "IS".

In order not to overburden Table 5, I consider in it, with respect to mathematical operations, only two options: the very operation of addition and a choice between the operations of addition and subtraction. The second option constrains the great ape less, and the level of constraint in Table 5 is the lowest in the right bottom cell of the second line.

The level of constraint would decrease even further if the great ape could, after mastering additional operations such as division and multiplication, choose an operation from the set $\{+, -, \div\}$ or even from the set $\{+, -, \div, \times\}$. The level of constraint would also decrease more and more with each increase of the number of natural numbers in the base.

It remains an empirical question, to be answered by means of experiments, which mathematical operations, besides addition, a great ape could master; if the great apes would be able to choose freely from a set involving at least two mathematical operations and what is the highest number of operations from which they would be able to choose. Another empirical question is if there is a limit to great apes' mastery of the number of natural numbers from the base.

In any case, *for the language of mathematics holds the same as for a non-mathematical language: the richer the language a great ape is taught by humans, the richer are the language capabilities of his or her own species, which can be manifested in language experiments with this great ape.*

References

Andrews, J. F., Logan, R. and Phelan, J. G. (2008): Milestones of language development. *Advance for Speech-Language Pathologists and Audiologists* 18: 1–20, 42.

Boysen, S. T. (1993): Counting in chimpanzees. In: Boysen, S. T., Capaldi, E. J. (Eds.): *The Development of Numerical Competence*. Hillsdale (NJ), Lawrence Erlbaum: 39–59.

Boysen, S. T., Berntson, G. G. (1989): Numerical competence in chimpanzee (*Pan troglodytes*). *Journal of Comparative Psychology* 103 (1): 23–31.

Brown, R. (1980): The first sentences of child and chimpanzee. In: Seboek, T. A., Umiker-Seboek, J. (Eds.): *Speaking of Apes*. New York, Plenum Press: 85–102.

Carnap, R. (1956): *Meaning and Necessity*. Chicago: Chicago University Press.

Cmorej, P. (1985): K explikácii pojmu jazykový znak [On the explication of the notion of language sign]. *Jazykovedný časopis* 36 (2): 150–162.

Cmorej, P. (1998): Denotácia a referencia [Denotation and reference]. In: Cmorej (Ed.): *K filozofii jazyka, vedy a iným problémom* [*On the Philosophy of Language, Science and other Problems*]. Bratislava, Infopress: 7–19.

Cmorej, P. (2001): *Úvod do logickej syntaxe a sémantiky* [*Introduction to Logical Syntax and Semantics*]. Bratislava: IRIS.

Dore, J. (1974): A pragmatic description of early language development. *Journal of Psycholinguistic Research* 3 (4): 343–350.

Dore, J. (1975): Holophrases, speech acts and language universals. *Journal of Child Language* 2 (1): 21–40.

Essock, S. M., Gill, T. V. and Rumbaugh, D. M. (1977): Language relevant object- and color-naming tasks. In: Rumbaugh (Ed.) 1977: 193–206.

Fields, W. M., Sagerdahl, P. and Savage-Rumbaugh, E. S. (2007): The material practices of ape language research. In: Valsinen, J., Rosa, A. (Eds.): *The Cambridge Handbook of Sociocultural Anthropology*. Cambridge, Cambridge University Press: 164–186.

Fouts, R. S. (1972): The use of guidance in teaching sign language to chimpanzee (*Pan troglodytes*). *Journal of Comparative and Physiological Psychology* 80 (3): 515–522.

Fouts, R. S. (1973): Acquisition and testing of gesture signs in four young chimpanzees. *Science* 180 (4089): 978–980.

Fouts, R. S. (1974a): Language. *Journal of Human Evolution* 3 (6): 475–482.

Fouts, R. S. (1974b): Talking with chimpanzees. In: *Science Year – World Book Science Annual 1974*. Chicago, Field Enterprises Educational Corporation: 34–49.

Fouts, R. S. (1975): Communication with chimpanzees. In: Eibl-Eibesfeldt, I., Kurth, G. (Hg.): *Hominization and Behavior*. Stuttgart, Fischer: 137–158.

Fouts, R. S. (1977): AMESLAN in Pan. In: Bourne, G. H. (Ed.): *Progress in Ape Research*. New York, Academic Press: 117–123.

Fouts, R. S. (1978): Sign language in chimpanzee. In: Peng, F. C. C. (Ed.): *Sign Language and Language Acquisition in Man and Ape*. Boulder (CO), Westview Press: 121–136.

Fouts, R. S., Fouts, D. H. (1993): Chimpanzees use of sign language. In: Cavalieri, P., Singer, P. (Eds.): *The Great Apes Project*. New York, St. Martin's Press: 28–41.

Fouts, R. S., Mellgren, R. L. (1976): Language intervention in ecological and ethological perspective. In: Button, J. E.,

94

Lovitt, T. C. and Rowland, R. D. (Eds.): *Communication Research in Learning Disabilities and Mental Retardation*. Baltimore, University Park Press: 249–281.

Fouts, R. S., Chown, B. and Gooding, L. (1976): Transfer of signed responses in American Sign Language from vocal English stimuli to physical object stimuli by chimpanzee (*Pan troglodytes*). *Learning and Motivation* 7 (3): 458–475.

Fouts, R. S., Fouts, D. H. and Schoenfeld, D. (1984): Sign language conversational interaction between chimpanzees. *Sign Language Studies* 42: 1–12.

Fouts, R. S., Fouts, D. H. and Van Cantfort, T. E. (1989): The infant Loulis learns signs from cross-fostered chimpanzees. In: Gardner, Gardner and Van Cantfort (Eds.) 1989: 280–292.

Fouts, R. S., Hirsch, A. D. and Fouts, D. H. (1982): Cultural transmission of a human language in a chimpanzee mother/infant relationship. In: Fitzgerald, H. E., Mullins, J. A. and Page, P. (Eds.): *Psychological Perspectives; Child Nurturance Series*, Vol. III. New York, Plenum Press: 159–193.

Fouts, R. S., Shapiro, G. and O'Niel, C. (1978): Studies of linguistic behavior in apes and children. In Siple, P. (Ed.): *Understanding Language through Sign Language Research*. New York, Academic Press: 163–185.

Gardner, B. T., Gardner, R. A. (1971): Two-way communication with an infant chimpanzee. In: Schrier, Stollnitz (Eds.) 1971: 117–184.

Gardner, B. T., Gardner, R. A. (1974): Comparing the early sentence utterances of child and chimpanzee. In: Bick, A. (Ed.): *Minnesota Symposia on Child Psychology*, Vol. 8. Minneapolis, University of Minnesota Press: 3–23.

Gardner, B. T., Gardner, R. A. (1975): Evidence for sentence constituents in the early utterances of child and chimpanzee. *Journal of Experimental Psychology – General* 104 (3): 244–267.

Gardner, B. T., Gardner, R. A. (1980): Two comparative psychologists look at language acquisition. In: Nelson, K. (Ed.): *Children Language*, Vol. 2. New York, Gardner Press: 331–369.

Gardner, B. T., Gardner, R. A. and Nichols, S. G. (1989): The shapes and uses of signs in a cross-fostering laboratory. In: Gardner, Gardner and Van Cantfort (Eds.) 1989: 55–180.

Gardner, B. T., Gardner, R. A. (1998): Development of phrases in the early utterances of children and cross-fostered chimpanzees. *Human Evolution* 13 (3–4): 161–188.

Gardner, R. A., Gardner, B. T. (1969): Teaching sign language to a chimpanzee. *Science* 165 (3894): 664–672.

Gardner, R. A., Gardner, B. T. (1978): Comparative psychology and language acquisition. *Annals of the New York Academy of Sciences* 309: 37–76.

Gardner, R. A., Gardner, B. T. (1989a): Cross-fostered chimpanzees: I. In: Heltne, P. G., Marquardt, L. A. (Eds.): *Understanding Chimpanzees*. Cambridge (MA), Harvard University Press: 224–233.

Gardner, R. A., Gardner, B. T. (1989b): A cross-fostering laboratory. In: Gardner, Gardner and Van Cantfort (Eds.) 1989: 1–28.

Gardner, R. A., Gardner, B. T. (1998): Ethological study of early language. *Human Evolution* 13 (3–4): 189–207.

Gardner, R. A., Gardner, B. T. and Van Cantfort, T. E. (Eds.) (1989): *Teaching Sign Language to Chimpanzee*. Albany (NY): SUNY Press.

Gardner, R. A., Van Cantfort, T. E. and Gardner, B. T. (1992): Categorical replies to categorical questions by cross-fostered chimpanzees. *American Journal of Psychology* 105 (1): 27–57.

Gill, T. V., Rumbaugh, D. M. (1974): Mastery of naming skills by a chimpanzee (*Pan*). *Journal of Human Evolution* 3 (2): 483–492.

Gill, T. V., Rumbaugh, D. M. (1977): Training strategy and tactics. In: Rumbaugh (Ed.) 1977: 157–192.

Glasersfeld, E. von (1974): The Yerkish language for non-human primates. *American Journal for Computational Linguistics* 1 (1): 1–56.

Glasersfeld, E. von (1977): The Yerkish language and its parser. In: Rumbaugh (Ed.) 1977: 91–130.

Glasersfeld, E. von *et al.* (1973): A computer mediates communication with a chimpanzee. *Computers and Automation* 22 (7): 9–11.

Habermas, J. (1983): *Theory of Communicative Action.* Vol. 1. Boston: Beacon Press.

Habermas, J. (1984): Aspekte der Handlungsrationalität [Aspects of the rationality of action]. In: Habermas, J.: *Vorstudien und Ergänzungen zur Theorie des kommunikativen Handelns* [*Preliminary Studies and Addenda to the Theory of Communicative Action*]. Frankfurt am Main, Suhrkamp: 441–472.

Habermas, J. (1988): *On the Logic of Social Sciences.* Cambridge (MA): MIT Press.

Habermas, J. (2001): *On the Pragmatics of Social Interaction.* Cambridge (MA): MIT Press.

Habermas, J. (2003): *Truth and Justification.* Cambridge (MA): MIT Press.

Jaffe, S., Jensvold, M. L. A. and Fouts, D. H. (2002): Chimpanzee to chimpanzee signed interactions. In: Landau, V. (Ed.): *Chimpanzoo Conference Proceedings*. Tucson (AZ): 67–75.

Jensvold, M. L. A., Fouts, R. S. (1993): Imaginary play in chimpanzees (*Pan troglodytes*). *Human Evolution* 8 (3): 217–227.

Jensvold, M. L. A., Gardner, R. A. (2000): Interactive use of sign language by cross-fostered chimpanzees. *Journal of Comparative Psychology* 114 (4): 335–346.

Leeds, C. A., Jensvold, M. L. A. (2013): Communicative functions of five signing chimpanzees (*Pan troglodytes*). *Pragmatics and Cognition* 21 (1): 224–247.

Lyn, H., Russell, J. L. and Hopkins, W. D. (2010): The impact of environment on the comprehension of declarative communication in apes. *Psychological Science* 21 (3): 360–365.

Lyn, H. *et al.* (2011): Nonhuman primates declare! *Language and communication* 31 (1): 63–74.

Matsuzawa, T. (1985): Use of numbers by a chimpanzee. *Nature* 315 (6014): 57–59.

Miles, L. W. (1976): The communicative competence of child and chimpanzee. *Annals of the New York Academy of Sciences* 280: 592–597.

Miles, L. W. (1983): Apes and language. In: Luce, J. de, Wilder, H. T. (Eds.): *Language in Primates*. New York, Springer: 43–61.

Miles, L. W. (1990): The cognitive foundations of reference in a signing orangutan. In: Parker, S. T., Gibson, K. R. (Eds.): *'Language' and Intelligence in Monkeys and Apes*. Cambridge, Cambridge University Press: 511–539.

Miles, L. W. (1994): ME CHANTEK. In: Parker, S. T., Mitchell, R. W. and Boccia, M. L. (Eds.): *Self-awareness in Animals and Humans*. Cambridge, Cambridge University Press: 254–272.

Premack, D. (1970a): A functional analysis of language. *Journal of the Experimental Analysis of Behavior* 14 (1): 107–125.

Premack, D. (1970b): Education of Sarah. *Psychology Today* 4 (4): 54–58

Premack, D. (1971a): Language in chimpanzee? *Science* 172 (3985): 808–822.

Premack, D. (1971b): On the assessment of language competences in chimpanzee. In: Schrier – Stollnitz (Eds.) 1971: 185–228.

Premack, D. (1971c): Some general characteristics of a method for teaching language to organisms that do not ordinarily acquire it. In: Jarrard, L. E. (Ed.): *Cognitive Processes of Nonhuman Primates*. New York, Academic Press: 47–82.

Premack, D. (1976): *Intelligence in Ape and Man*. Hillsdale (NJ): Lawrence Erlbaum.

Premack, D. (1986): *Gavagai!* Cambridge (MA): MIT Press.

Rivas, E. (2005): Recent use of signs by chimpanzees (*Pan troglodytes*) in interactions with humans. *Journal of Comparative Psychology* 119 (4): 404–417.

Rumbaugh, D. M. (Ed.) (1977): *Language Learning by a Chimpanzee*. New York: Academic Press.

Rumbaugh, D. M., Gill, T. V. (1976): Language and the acquisition of language-type skills by a chimpanzee (*Pan*). *Annals of the New York Academy of Sciences* 270: 90–123.

Rumbaugh, D. M., Gill, T. V. (1977): Lana's acquisition of language skills. In: Rumbaugh (Ed.) 1977: 165–192.

Rumbaugh, D. M., Savage-Rumbaugh, E. S. (1978): Chimpanzee language research. *Behavior Research Methods and Instrumentation* 10 (2): 119–131.

Rumbaugh D. M., Savage-Rumbaugh E. S. and Hegel, M. T. (1987): Summation in chimpanzee (*Pan troglodytes*). *Journal of Experimental Psychology – Animal Behavioral Process* 13 (2): 107–115.

Rumbaugh D. M., Savage-Rumbaugh E. S. and Pate, J. L. (1988): Addendum to "Summation in chimpanzee (*Pan troglodytes*)." *Journal of Experimental Psychology – Animal Behavioral Process* 14 (1): 118–120.

Rumbaugh, D. M., Warner, H. and Glasersfeld, E. von (1977): The LANA project. In: Rumbaugh (Ed.) 1977: 87–90.

Rumbaugh, D. M. *et al.* (1973): A computer-controlled language training system for investigating the language skills of young apes. *Behavior Research Methods and Instrumentation* 5 (5): 385–392.

Rumbaugh, D. M. *et al.* (1975): Conversations with a chimpanzee in computer-controlled environment. *Biological Psychiatry* 10 (6): 627–641.

Savage-Rumbaugh, E. S. (1981): Can apes use symbols to represent their world? *Annals of the New York Academy of Sciences* 364: 35–59.

Savage-Rumbaugh, E. S. (1984a): Acquisition of functional symbol usage in apes and children. In: Roitblat, H. L., Bever, T. G. and Terrace, H. S. (Eds.): *Animal Cognition.* Hillsdale (NJ), Lawrence Erlbaum: 291–310.

Savage-Rumbaugh, E. S. (1984b): Verbal behavior at the procedural level. *Journal of the Experimental Analysis of Behavior* 41 (2): 223–250.

Savage-Rumbaugh, E. S. (1987): Communication, symbolic communication, and language. *Journal of Experimental Psychology – General* 116 (3): 288–292.

100

Savage-Rumbaugh, E. S. (1999): Ape Language. In: King, J. B. (Ed.): *The Origins of Language*. Santa Fe, School of American Research Press: 115–188.

Savage-Rumbaugh, E. S., Fields, W. M. (2006): Rules and Tools. In: Toth, N., Schick, K. (Eds.): *The Oldowan*. Bloomington (IN), The Stone Age Institute: 223–242.

Savage-Rumbaugh, E. S., Rumbaugh, D. M. and Boysen, S. (1978a): Linguistically mediated toll use and exchange by chimpanzees (*Pan troglodytes*). *Behavior and Brain Sciences* 1 (4): 539–554.

Savage-Rumbaugh, E. S., Rumbaugh, D. M. and Boysen, S. (1978b): Symbolic communication between two chimpanzees. *Science* 201 (4356): 641–644.

Savage-Rumbaugh, E. S., Rumbaugh, D. M. and Boysen, S. T. (1980): Do apes use language? *American Scientist* 68 (1–2): 49–61.

Savage-Rumbaugh, E. S., Rumbaugh, D. M. and McDonald, K. (1985): Language learning in two species of apes. *Neuroscience and Behavioral Reviews* 9 (4): 653–665.

Savage-Rumbaugh, E. S. *et al.* (1980): Reference. *Science* 210 (4472): 922–925.

Savage-Rumbaugh, E. S. *et al.* (1986): Spontaneous symbol acquisition and communicative use by pigmy chimpanzees (*Pan paniscus*). *Journal of Experimental Psychology – General* 115 (3): 211–236.

Savage-Rumbaugh, E. S. *et al.* (2005): Culture prefigures cognition in *Pan/Homo* bonobos. *Theoria* 20 (3): 311–328.

Schrier, A. M., Stollnitz, F. (Eds.) (1971): *Behavior of Nonhuman Primates*, Vol. 4. New York: Academic Press.

Searle, J. R. (1969): *Speech Acts*. Cambridge: Cambridge University Press.

Searle, J. R. (1975): A taxonomy of illocutionary acts. In: Gunderson, K. (Ed.): *Minnesota Studies in the Philosophy of Science* 7. Minneapolis, University of Minnesota Press: 344–369.

Searle, J. R. (2001): Meaning, mind and reality. *Revue internationale de philosophie* 55 (216): 173–179.

Stokoe, W. C., Jr. (1960): Sign language structure. *Studies in Linguistics, Occasional Papers* 8: 1–78.

Stokoe, W. C., Jr. (1972): *Semiotics and Human Sign Languages*. The Hague: Mouton.

Tichý, P. (1995): Constructions as the subject-matter of mathematics. In: Depauli-Schimanovisch, W., Köhler, E. and Stadler, F. (Eds.): *The Foundational Debate*. Reidel, Dordrecht: 175–185.

Zouhar, M. (2002a): Referencia vlastných mien [Reference of proper names] (I). *Organon F* 9 (3): 272–293.

Zouhar, M. (2002b): Referencia vlastných mien [Reference of proper names] (II). *Organon F* 9 (4): 338–357.

Zouhar, M. (2003): Referencia vlastných mien [Reference of proper names] (III). *Organon F* 10 (1): 18–48.

www.ingramcontent.com/pod-product-compliance
Lightning Source LLC
Chambersburg PA
CBHW071750270326
41928CB00013B/2860